Design and Analysis of Cold-formed Sections

Design and Analysis
of
Cold-formed Sections

edited by A. C. WALKER

A HALSTED PRESS BOOK

JOHN WILEY & SONS
New York—Toronto

First published by
International Textbook Company Limited
Kingswood House, Heath & Reach, Leighton Buzzard, Beds. LU7 OAZ
and 450 Edgware Road, London W2E 1EG
in association with the Cold Rolled Sections Association

First published 1975

Published in the USA and Canada by Halsted Press, a Division of
John Wiley & Sons, Inc., New York

Library of Congress Cataloging in Publication Data

Walker, A. C.
    Design and analysis of cold-formed sections.
    "A Halsted Press book."
    Includes bibliographical references and indexes.
    1. Steel, Structural.   2. Structural design.
3. Structural stability.   I. Title.
TA684.W28   624'.182   75-1315
ISBN 0-470-91809-8

Printed in Great Britain

Foreword

The field of 'cold-formed' structures — which are structural shapes formed from metal sheet and strip in the cold state — is one of growing importance in structural engineering and building. Structural materials will inevitably become more expensive, and there will be increasing pressures to use thinner metallic structural members to save material. In this drive for economy, the cold-formed structural section has a crucial role to play, and it is very appropriate that the scientific basis of the structural use of such sections should be spelt out clearly and effectively.

Dr Alastair Walker, with his colleagues in industry and the universities, brings together the fundamental basis of structural design of these sections in this important text. It is based on the results of many years of painstaking study of design problems and the research needed to solve these. The outcome is a very comprehensive treatment of the structural design of these sections, and the book should become a standard international text in its field.

In Britain, much of the work on which these results are based has been supported over many years by the Cold Rolled Sections Association of Great Britain; in sponsoring this work the Association has made an important contribution to the development of structural design in cold-formed sections.

I am sure this book will become an important source of reference. It will benefit not only the economical and rational use of structural materials, but also advance our knowledge of structural design in the important direction of thinner structural sections.

A. H. Chilver
Cranfield Institute of Technology

List of Contributors

Dr M. M. Black,*   Reader in Structural Engineering, University of Sussex

Dr R. G. Dawson,**   Managing Director of Bromwich Structures, Ltd

Professor J. M. Harvey,* Professor of Mechanics of Materials, Strathclyde University

Dr P. J. B. Morrell,   University of Sussex

Dr J. Rhodes,   Strathclyde University

Dr A. C. Walker,*   Reader in Structural Engineering, University College, London

* Technical Adviser to the Cold Rolled Sections Association
** Member of the Technical Committee of the Cold Rolled Sections Association

Cold Rolled Sections Association List of Members

AYRSHIRE METAL PRODUCTS LTD

Church Street,
Irvine, Ayrshire,
Scotland KA2 8PH

BRITISH STEEL CORPORATION
STRIP MILLS DIVISION

Godins Works, P. O. Box 28,
Mendalgief Road, Newport,
NPT 2WX

BRITISH URALITE LTD

Higham,
Rochester, Kent

COLD ROLLED SECTIONS LTD

Albion Road,
West Bromwich,
Staffs WV13 8SW

DUCTILE SECTIONS LTD

Planetary Road,
Willenhall, Staffs

EASTERN SECTIONS LTD

Ayton Road, Wymondham,
Norfolk

ELWELL SECTIONS LTD

Phoenix Street,
West Bromwich,
Staffs B70 0AQ

STEWART FRASER LTD

Henwood Industrial Estate,
Ashford, Kent TN24 8DR

METAL MOULDINGS LTD

Park Royal Road,
London NW10 7LP

METAL SECTIONS LTD

Broadwell Works, Oldbury,
Warley, Worcs

METAL TRIM LTD	Royal Oak Way, Daventry, Northants
ROLLER SHUTTERS LTD	Anne Road, Smethwick Warley, Worcs
SAFCON (STRATFORD) LTD	Cooks Road, London E15 2PN
WARWICK RIM & SECTIONING CO. LTD	Bagnall Street, Golds Green, West Bromwich, Staffs

SECRETARIES

ROBSON RHODES	King Edward House, New Street, Birmingham B2 4QP

Contents

Preface

This book provides an introduction to the analysis and design of cold-formed structural components. Also, the intention is to form a bridge between the analyses presented in technical papers and mathematically oriented textbooks and the practical design methods that have been distilled from such analyses. The text therefore refers often to the British Standard 449 Addendum No. 1 (PD 4064) which is the 'Specification for the Use of Cold Formed Steel Sections in Building' and indicates the sources and bases of the design recommendations in that document. Wherever possible, experimental results are presented to enable the reader to judge the safety of the various design formulations. Examples of the use of the analytical and design approaches appear frequently in the text.

In the presentation of the various analytical methods applicable to cold-formed components it has been assumed that the reader is familiar with the more conventional methods taught in undergraduate courses and usually applied to the analysis of hot-rolled steel structural sections. Emphasis is therefore placed on those aspects of analysis that are peculiar to the design of thin-walled structures and include local buckling, warping torsion and torsional instability.

The first chapter gives a general introduction to the subject of the book and discusses the various methods of manufacture and the range of steels commonly used. Some areas of present application of cold-formed sections are indicated and the economic factors influencing their use are outlined. The chapter closes with an indication of the freedom of choice that is open to a designer when he exploits the capability of manufacturers to make almost any section geometry he cares to specify.

The design of columns is covered in Chapter 2 and includes a discussion on overall buckling, local buckling and torsional—flexural buckling. The particular feature of behaviour that arises from the thinness of cold-formed structural elements is of course local buckling. This is introduced in Chapter 2 and although a rigorous mathematical analysis, which is very complicated, is avoided, nevertheless this type of behaviour is treated very fully from a phenomenological and design viewpoint. The manner in which local buckling influences the overall buckling of thin-walled struts is discussed and the basis for a simplified design formulation is presented.

Chapters 3 and 4 deal with the design of thin-walled beams. Chapter 3 opens with a review of the engineer's theory of bending for unsymmetric sections. Again due to the thinness of the components, local buckling can occur and the engineer's theory of bending must be modified. In this chapter a simple design approach is presented which accommodates local buckling and which predicts stresses and deflections with an accuracy quite sufficient for engineering purposes. Buckling, this time due to the shear loads of the beam webs, is shown to be also a limiting factor on the load bearing capacity of cold-formed sections. Although this tends to be a difficult analytical problem when dealt with on a rigorous basis the approach in this book is to derive an empirical design formulation.

Torsion of course exists as a form of loading for all types of structural element. But as a result of the low torsional stiffness of thin-walled cold-formed sections the stresses and deformations that are set up by torques are much larger than in their hot-rolled counterparts. The various theories of torsion are reviewed in Chapter 4 and an attempt is made to simplify the design calculations for warping-torsion by making use of an analogy with beam flexure and by presenting the results of the rigorous mathematical theory in graphical form. The chapter closes with a discussion of the stress distribution in a continuous beam due to the application of torque loading.

In Chapter 5 we outline the problems of localized loading, and the methods of jointing. This includes the effect that buckling in the region of concentrated loads can have on the determination of the limiting load carrying capacity of cold-formed sections. The design method for beams presented in Chapter 3 is here extended to deal with troughed sheeting subject to various forms of loading.

This book comprises contributions from a number of people actively engaged in the research and design of cold-formed structures. Broadly the contributions took the form; Chapter 1 from Dr R. G. Dawson; Chapter 2 and part of Chapter 5 from Dr A. C. Walker; Chapter 3 and most of Chapter 5 from Professor J. M. Harvey and Dr J. Rhodes; Chapter 4 from Dr M. M. Black and Dr P. J. B. Morrell.

The editor is grateful to the contributors for their very valuable comments on the book at the draft stage. Also, he would like to acknowledge the help given to him by the members of the Technical Committee of the Cold Rolled Sections Association.

1

Introduction

The use of cold-rolled steel sections in structures represents one further step in the evolutionary development of structural form from its early massiveness to the present day's trend towards as slender a structure as possible.

Until the introduction of metals into structural engineering in the late eighteenth century, the majority of structures comprised stout stone columns and massive arches, although some attempts had been made to build thin masonry domes. Timber provided a possible exception but the use of this material was limited to smaller scale structures. These massive forms persisted initially with the introduction of metals — cast iron columns for example were of heavy, solid cross-section.

The potential economic gains of reducing the metal content of a structural member were at once obvious and have long been major factors dictating the development of structural forms to achieve as slender a shape as possible. The use of relatively thin metal elements in structures was first realized in the early nineteenth century with the development of plate girders and tubular bridges. The plate components of these were not thin by today's standards, but they represented in their time a very real attempt at making use of the high structural efficiency of thin structures as opposed to thick or massive structural forms.

Perhaps the most important development towards thin metal structures was the introduction of mild steel during the latter part of the nineteenth century. This led to the manufacture of large thin plates of a strong ductile material which could be connected to other components either by riveting or bolting, and enabled established practice in wrought iron tubular structures and deep web beams to be consolidated and extended. A whole new field of structures, on both the large and small scales, was brought about and, except for the introduction of high strength materials and welded connections, it still provides much of the present-day basis of the field of thin structures.

With the development of the aeroplane in the early twentieth century came

1

an overriding need for as light a structural form as possible. This was usually achieved by using thin metal sheets to which stiffeners (sometimes called stringers) were riveted. These stiffeners were usually of extruded aluminium channel, Z or H section. Much of the early theoretical and empirical analysis of the behaviour of thin gauge structures as we know it today followed the development of this structural form. Because of the weight savings offered, the application to aircraft was mainly in aluminium, but the principles evolved are equally applicable to steel. As with much structural development, theoretical analysis has tended to follow practical application and the use of cold-rolled sections is no exception.

The next stage in the development of cold-formed steel structural elements came not in the field of structures but in the automobile industry. During the First World War, techniques were developed whereby body sections of cars and lorries etc. could be produced by bending or shaping light gauge sheet, in the cold state, continuously along its length. These techniques were then applied to produce window frames and similar non-structural elements of a building. From this, and with the experience being gained in the aircraft industry, it was but a short step to the production of thin steel shapes which could be used structurally in a building — the era of the cold-formed structural steel element was born. Because of the shortage of steel during the Second World War, the advantages of saving weight were obvious and the use of thin structural elements was consolidated. Since then this use has flourished, at first mainly in the USA, followed closely by this country and recently by many others throughout the world.

However, possibly because of conservatism and suspicion due to lack of understanding of the phenomena involved, the structural use of cold-formed steel sections is not as wide as it might be. It is with the aim of helping engineers and architects understand the behaviour of such structural elements that this book has been written.

1.1 SHAPES PRODUCED BY COLD FORMING

A very wide variety of shapes can be produced by cold forming, as illustrated in Figure 1.1, but generally it is only the simpler shapes that are used structurally. The dimensional and cross-sectional properties of a typical range of such shapes are given in a British Standard, BS 2994. However, this document does not represent the total range nor even necessarily the most easily obtained sizes, for most manufacturers roll their own sections on a proprietary basis for specific purposes.

The commonest sections are Zeds, angles and channels which may be plain or which may have 'lips' added. These lips are small additional elements provided to a section to improve its efficiency under compressive loads. The action of a lip is described fully in Chapter 2 of this book. Structural efficiency can also be improved by welding two channels back to back to form an I section — the advantage

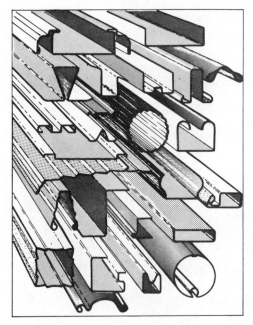

Figure 1.1

of this being that the natural disposition for a channel section to twist under bending loads is overcome. Again this is amplified later in this book.

Other popular cold-formed sections are the different trough-shaped elements which are now widely used for roof decking and wall cladding on many industrial buildings. These are usually provided as pre-galvanized or plastic coated and need minimal maintenance throughout their working life.

1.2 GENERAL APPLICATIONS

The applications to which cold-formed sections are put must be nearly as diverse as the types of sections produced; some examples of the types of structure embodying cold-formed sections are illustrated in Figures 1.2–1.5.

1.2.1 PURLINS
In general sections are used structurally in less heavily loaded situations where common hot-rolled sections would be understressed and therefore not used to full efficiency. Common examples of such use are Z-shaped sections instead of

Figure 1.2   The use of Z-purlins in an industrial building

hot-rolled angle for purlins in shallow sloped roofs of industrial building and the
boom sections of lattice beams used widely as joists to floors and flat roofs. The
relative economics of using Z purlins instead of hot-rolled sections are shown in
Figure 1.6. This shows that, for the very common typical case chosen, cold-
rolled Z's are more economical than hot-rolled sections over the approximate
span range 3.75 − 9 m. As this covers virtually the whole of this type of construction the usage of Z's is very widespread.

    Z sections are used rather than channels for a variety of reasons, the main one
being that their principal axis often coincides with the roof pitch thus enabling
the designer to take full advantage of the intrinsic strength of this section. A
secondary advantage is they offer easier fixing − the other flange doesn't get in
the way of the spanners etc. The sections are rolled in several different depths
and thicknesses to cover a wide range of load span combinations. The sections
are usually supplied continuously over two or three spans − except where this
would mean transporting lengths greater than 18 m which requires special legis-
lation − so incurring disproportionate costs. Adjacent lengths are generally joined
over a rafter back usually using a special sleeve of the same shape as the purlin.
This offers a fair measure of continuity over the joint so helping to distribute

Figure 1.3   Cold-formed section used as the framing for a bus body

moment. The purlins are connected to the rafter using substantial cleats which provide torsional restraint to the section which would otherwise twist. It is also important that a suitable connection is effected between the roof sheeting and the purlins so that the interaction between the two restrains the Z from twisting at mid-span. Experience has shown that with asbestos cement sheets properly connected hook bolts at about 1m centres produce the interaction required. Self-tapping or shot-fired fixings through the valleys of a metal decking have been found to serve the same purpose and also to offer the roof a significant degree of in-plane stiffness. Fixings through the top of the corrugation are not considered as favourably because the shear stiffness so offered by the decking is less than that for fixing through the 'valleys'.

For longer spans, sag bars are used, partly to assist during sheeting before the sheets have been fixed down, but also to help during wind uplift. There is some controversy as to how these sag bars should be fitted. It would seem to depend on the roof slope and direction of loading; at low slopes the Z — which always points 'up' the slope — twists 'forwards' and therefore should be restrained 'top to bottom', as in Figure 1.7(a); at higher slopes the Z twists 'backwards' so that better restraint is provided if the sag bars are fitted 'bottom to top', as in Figure

Figure 1.4   Cold formed channel beams and struts

1.7(b). It may be necessary to modify this approach if wind loading is critical in design.

Z sections have also been used as side rails to support the vertical cladding to a building. Great care has to be exercised in this instance to minimize twist under dead load, and double banked sag rods from a stiff eaves beam are recommended. As an alternative to the Z section a 'sigma' profile (Σ) has been introduced in the UK following successful application in the United States. The advantages claimed

Figure 1.5 Large building constructed entirely from cold formed section. Note the use of lattice joist purlins

for this section are that its shape is such as to bring the shear centre within the cross-section and so considerably to reduce the tendency to twist. The reasons for this are discussed fully in Chapter 4.

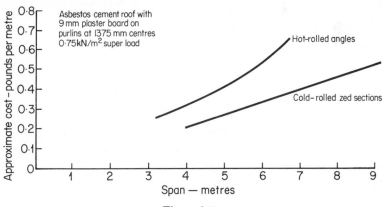

Figure 1.6

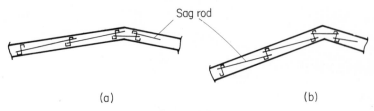

Sag rod

(a)   (b)

Figure 1.7

1.2.2 LATTICE BEAMS

Open web lattice beams are made using several different boom shapes, from plain or lipped angle sections to the more complicated members such as those shown in Figure 1.8. The different elements are rolled in different sizes and thickness and then welded to diagonals formed from continuous lengths of round mild steel bar bent to the required pattern. In this way beams of different strength can be made in depths up to 1·8m or so, offering a wide range of strengths for different purposes. The beams can easily be precambered against dead load so that, in those cases where this is a significant proportion of the total load, considerable increases in the allowable design load can be obtained. The weights of these beams range between about 5 and 60kg/m, although more recently much lighter beams down to about 1.2kg/m have been produced and are proving competitive in the domestic housing market as substitutes for conventional timber joists.

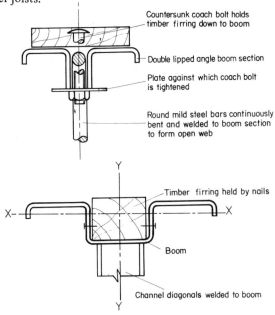

Countersunk coach bolt holds timber firring down to boom

Double lipped angle boom section

Plate against which coach bolt is tightened

Round mild steel bars continuously bent and welded to boom section to form open web

Timber firring held by nails

Boom

Channel diagonals welded to boom

Figure 1.8

The relative economics of using these lightweight lattice beams to support a typical flat roof construction, such as those used widely in schools, hospitals, warehouses etc., are illustrated in Figure 1.9. In this case the cost advantages of using the lattice joist cover the approximate span range 4.5—15 m. Obviously the curves shown in Figure 1.9 cannot be taken as exact because cost depends on a number of indeterminates such as quantity, delivery, special fixings, section

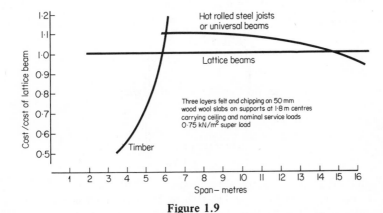

Figure 1.9

difficulties etc. Also the curves should really be stepped between available sections, but here these steps have been smoothed out. Nevertheless, because of the cost savings suggested by typical cases such as are shown in Figure 1.9, lightweight open web joists are finding increasing popularity in standard building systems for classrooms, office blocks and the like up to two and sometimes even three storeys high. These systems use prefabricated timber wall, floor, ceiling and roof panels, thus allowing very rapid rates of construction on site. With the increasing cost of site labour this can offer significant cost benefits to the client. A further advantage of these lightweight open web joists is the freedom with which services can be passed within roof or floor zones.

1.2.3 OTHER APPLICATIONS
The main benefits of using a cold-formed section are not only its high strength to weight ratio but also its lightness — so saving on transport, erection and foundation costs — and the ability to distribute the material in the section to where it is most useful. In this way both strength and functional needs can be served. For example, the lattice beam section shown in Figure 1.8 offers good geometric properties about both axes, easy connection for the diagonals, and a facility for including a timber insert to which roof decking can be nailed.

Similarly two 'top hat' sections can easily be spot welded together to produce a highly efficient box section (Figure 1.10) to which bracing or wall panels etc.

can readily be attached. Such sections are used for the columns of industrialized building systems where holes for connections can be punched at standard positions. Other applications for 'line' elements include the members of lightweight lattice portals and the framing members for trailers, caravans, railway wagons and the like. Most of the slotted-angle sections on the market are cold-rolled; these are used for a multitude of purposes, ranging from both light and heavy racking to the complete structural framework for sheds and houses (on the overseas markets).

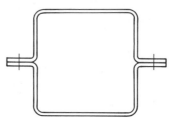

Figure 1.10

The other types of cold-formed section are the 'spatial' elements — those in which the breadth is of the same order as the length. These include the previously mentioned troughed sheets used for roof and wall cladding. Essentially the troughed sheet is used in two forms: as cladding, when it forms the outer skin, and as decking, when it forms (usually) the inner skin. Cladding can be to either roof or wall and is often either pre-galvanized or plastic coated for weather protection (the latter also offers a pleasing appearance). Thermal insulation is usually provided with an inner layer of foil backed plasterboard or similar material with or without an intermediate layer of glass fibre quilting. Decking is also used on roofs, the outer layer being made up of either asphalt or bituminous felt and stone chippings on an insulation board.

In some cases deeper troughed sheets have been used as floor decking. The latter application has been extended to produce composite concrete/steel floors which may be either prefabricated or have the concrete cast on site, in which case the steel decking is used as the form work. Experiments are continuing in this field to find improved methods of providing shear connection so that complete integral action is generated. Also of increasing interest are the savings that can be made in the cost of the supporting structure by using the in-plane stiffness provided by wall and roof sheeting to help transmit wind actions.[1]

1.3 COLD FORMING

There are two main ways of cold forming steel sheet, by *rolling* or by *press braking*. There are other ways, by drawing for example or by extrusion, but these are not economical in the mass production of prismatic steel sections and are therefore not widely used.

If sufficient quantities of a product are wanted, the coils may be delivered from the mills in the correct width necessary to form that product. More often though the strip is supplied from the mills in 'wide coil' form — usually in standard widths of 1200mm. This wide coil has then to be slit to the correct width — referred to as the *developed strip width* — to form the section under consideration. Slitting is carried out as a continuous operation using roller shears, the narrow strips being recoiled prior to forming.

1.3.1 COLD ROLLING

The process of cold rolling is illustrated schematically in Figure 1.11. The strip is formed gradually by feeding it continuously through successive pairs of rolls which act as male and female dies. Each pair of rolls progressively forms the strip until the finished cross-section is produced. The number of rolls, called 'stages', or sometimes 'passes', that are used depends on the complexity of the shape being rolled and the thickness and strength of the strip used. For example, five stages are usually sufficient to form a simple angle, whereas some complicated sections may require 15 or more stages.

The five or more stages needed to form a lipped Z shaped purlin progressively in thicknesses up to about 2mm ($\approx$ 14 gauge) are illustrated in Figure 1.12. Heavier gauges would need more passes, for example a 4mm thick Z ($\approx$ 8 gauge) would not be rolled on less than an eight stage mill and perhaps even a twelve stage. If the strip is formed too quickly, i.e. too few stages are used, the resultant section is liable to distortion both longitudinally, evidenced as twist, and of the cross-section. This latter effect is often referred to as *spring back*, and it occurs when the residual stresses due to rolling are released when the section is cut, causing some bends to open out and others to close in. Cutting is usually carried out using either a shear or a friction saw. The latter may be either static, the mill being stopped for each cutting operation, or it may be arranged to travel with the section so allowing much higher production rates to be achieved.

The principal advantage of cold roll forming, as compared with other methods of fabrication, is high production capacity. Speeds as high as 100m/min are reported for the forming of some products, but usual speeds are more like a quarter of this value, particularly for heavy sections. Production rates (i.e. including cutting to length, stacking, coil changes etc.) of between 600 and 1500m per working hour can be readily achieved in many cases.

Parts produced by roll forming are essentially uniform in cross-section and can be held within very close dimensional tolerances. Although tolerances of the finished part are dependent on the dimensional tolerances of the material being formed, in normal production a part should be held to within ± 0·4mm, though in some instances a closer tolerance is practicable on small shapes formed from

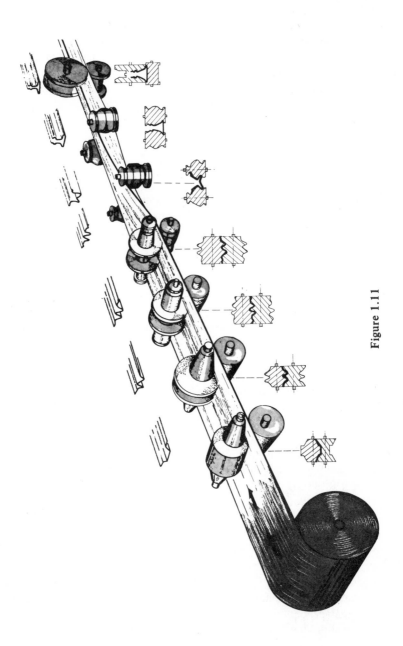

Figure 1.11

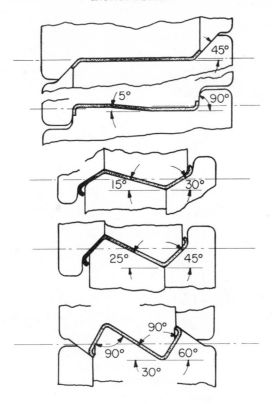

Figure 1.12

light gauge material. Another very important advantage of cold rolling is the ability to maintain fine surface finishes during roll forming operations. For example, painted and electroplated materials can be formed without damage to the coating. This is important with the pre-galvanized and plastic coated steels which are becoming increasingly popular.

1.3.2 PRESS BRAKING
The second method of forming sections in common use is by press braking. In this process, short lengths of strip are fed into the brake and pressed round shaped dies to form the final shape. This is illustrated schematically in Figure 1.13. Usually each bend is formed separately and the complexity of shape is limited to that into which the die can fit. In general, press beds are limited by power to lengths up to about 3 m, although some of the more powerful machines can form sections up to 8 m long.

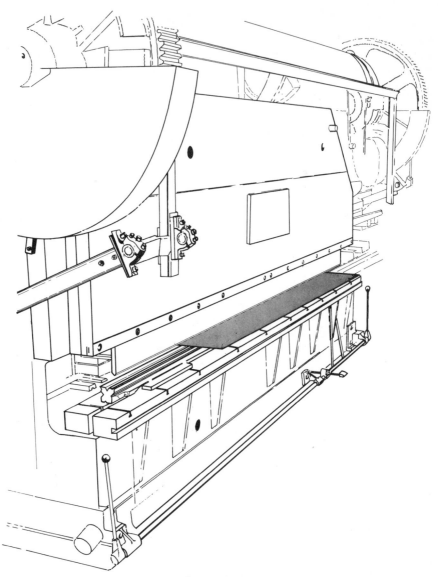

Figure 1.13

Whether to roll or brake depends largely on economics; in general the greater the quantity of a given section required the cheaper it is to form by rolling. This is because making a set of rolls to produce a section is very costly, and even if rolls already exist the expense of installing and commissioning them in a mill is a significant factor in the final cost of the section unless defrayed against considerable footage.

Figures 1.14, 1.15 and 1.16 illustrate the relative economics of roll forming and press braking three successively complicated sections.[2] Two curves each are shown for both rolling and pressing. One curve applies when tooling already

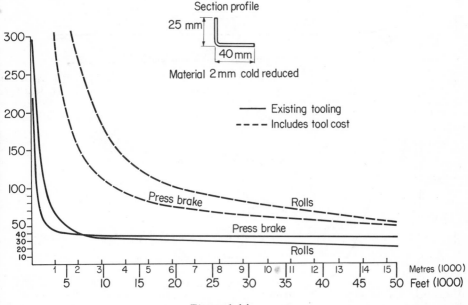

Figure 1.14

exists, and the second when new tools have to be commissioned and have to be amortized over the quantity of section to be supplied. The costs of production are stated as an 'on cost' to the basic cost of material; this is only a very rough parameter since of course costs change but it serves to illustrate the point. It can be seen that, using existing tooling, the break even point between rolling occurs between 1000 and 2500 m, but that if new tools are required this increases to between 10 000 and 20 000 m.

1.4 STEELS USED FOR COLD FORMING

Steels from which cold-formed sections are made are hot rolled by the mills to produce strip in uniform thicknesses of up to 8 mm. The preferred thicknesses

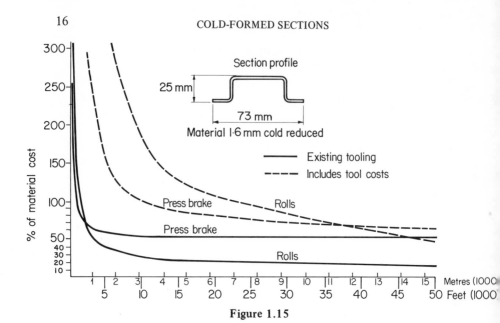

Figure 1.15

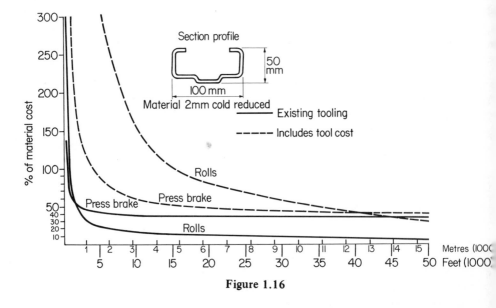

Figure 1.16

currently recommended by the British Standards Institution are listed in their document DD5. A comparison of these thicknesses with the old standard gauge sizes is given in Table 1.1. The steel is delivered in coil form, each coil containing possibly several hundred metres of material. The properties of steel produced in the UK are defined in BS 1449 part 1. *Specification for Steel Plate, Sheet and Strip*. The definitions of these terms are not absolutely clear but by common understanding:

> *Plate* is any material greater than or equal to 600mm wide and greater than or equal to 3 mm thick
> *Sheet* is any material greater than or equal to 600mm wide and less than 3mm thick
> *Strip* is all material less than 600mm wide.

BS 1449 lists a number of different steels depending on the purpose to which they are to be put. For example deep drawing steels, such as those used for making car body components, require properties quite different from those used to produce structural sections. In general the strength and ductility of a steel depend primarily upon its carbon content — the higher (within limits) the carbon, the stronger the steel but the lower its ductility.

Table 1.1

Standard steel sheet thickness – mm

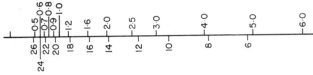

Imperial standard wire gauge

Gauge No.	26	24	22	20	18	16	14	12	10	8	6
Thickness-in	0·018	0·022	0·028	0·036	0·048	0·064	0·080	0·104	0·128	0·160	0·192
—mm	0·457	0·599	0·711	0·914	1·219	1·626	2·032	2·642	3·251	4·064	4·877

The properties of the steels generally used to produce structural sections are summarized in Table 1.2. Alternatively, higher strength steel can be produced by *cold reducing*, that is by 'squeezing' the strip between rollers to reduce its thickness. This 'cold work' on the steel improves its tensile and yield strengths but lowers its ductility. For example a 20% reduction in thickness can increase yield strength by 50% but reduces elongation to as little as 7%, which probably represents the limit of formability for simple shapes.

The relative economics of using these very different grades of steel for two typical strut sections are shown in Figures 1.17 and 1.18. In these figures the cost to support 1 kN in 43/25 grade steel has been taken as the base. Provisional allowance has been made for additional rolling costs of the harder steel, but no

COLD-FORMED SECTIONS

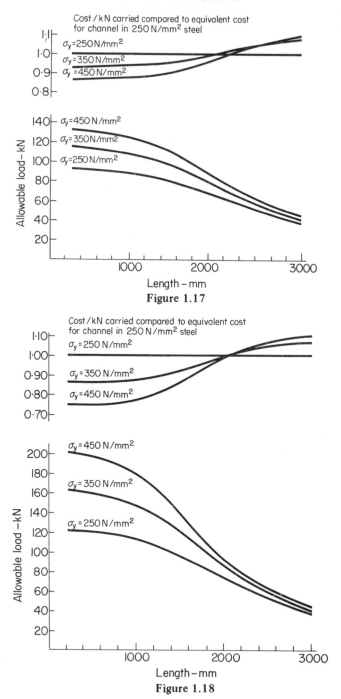

Figure 1.17

Figure 1.18

Table 1.2

Material Grade	Maximum Carbon Content %	Minimum Ultimate Tensile Strength N/mm^2	Minimum Yield Stress N/mm^2	Minimum Elongation (L_0 = 50mm) %
34/20	0.15	340	200	29
37/23	0.20	370	230	28
43/25	0.25	430	250	25
43/28	0.25	430	280	26

allowance has been made for problems of supply etc. As can be seen, the higher grade steels are more efficient (strength per unit cost) at shorter lengths but less so at longer ones. The situation is similar for beam sections; in the range in which stress is the limiting factor the higher grade steels will prove economic, but no advantage is offered where deflection is critical.

Higher strength steels can also be produced by the inclusion of small quantities of niobium and manganese, but these are usually more expensive. All the steels mentioned above are weldable provided that the welding techniques used make allowance for composition and thickness.

Typical stress/strain characteristics for these different grades for steel are shown in Figure 1.19. Note that there is no sharply defined yield point, and the

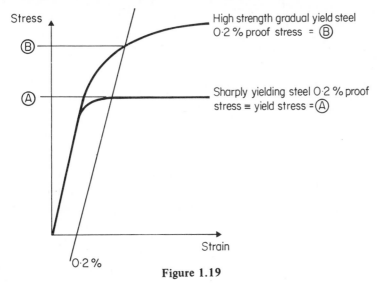

Figure 1.19

yield stress for structural purposes is taken as the 0.2% proof stress. Tests have shown that there is some variation in the value of Young's modulus of these steels, especially if they have undergone any degree of cold work. For the design examples in this book we have used the value of 200kN/mm^2.

One of the important characteristics of a cold formed product is that very often the basic material, that is the strip, has a better surface finish than can be obtained for 'heavy' hot rolled sections. A further improvement in the surface finish can be achieved by 'pickling' the strip to remove surface scale. If a very fine surface finish is required the strip may be pickled and then be given a 'skin pass' by passing it through a set of rollers which apply slight pressure.

1.5 WORK HARDENING DUE TO COLD FORMING

Because of the mechanical work done on forming a bend in a section, the strength of the material in the proximity of the bend is raised (Figure 1.20).[3] This strain hardening increases both the yield and ultimate stress (the increase in the

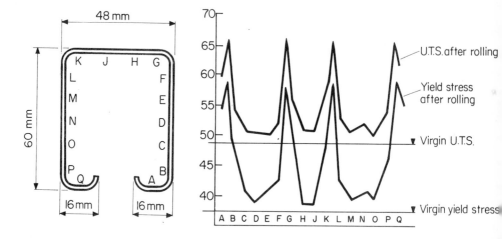

Figure 1.20

yield being greater than that in the ultimate) but reduces ductility. The degree and extent of the strain hardening seems to depend on a number of factors such as the method of forming, the quality of the virgin steel, the thickness of the steel and the tightness of the bend. As such the phenomenon cannot be accurately quantified and is not taken into account in the British design specification.[4]

However, American research[5] has shown that the order of increase can be worth utilizing, but that this should only be done for individual sections on a basis of statistically interpreted test results.

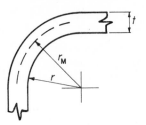

Figure 1.21

The work done on the bend not only strain hardens the material, it also causes some thinning as well. This means that there is a slight inward shift of the mid-plane round the bend. Kaltprofile[6] suggests that this shift can be calculated from

$$r_m = r + \kappa t \tag{1.1}$$

in which (see Figure 1.21) r_m is the mid-plane radius, r is the internal bend radius, t is the thickness of the plate and κ is given in Table 1.3.

In BS 2994 the method is modified slightly to

$$r_m = r + 0.5 t_{red} \tag{1.2}$$

where

$$t_{red} \equiv \left(\frac{r + \kappa t}{r + 0.5 t} \right) t$$

and

$$\kappa = 0.3 \text{ for } r/t = 1$$

$$\kappa = 0.35 \text{ for } r/t = 1.5$$

Table 1.3

r/t	>0.65	>1.0	>1.5	>2.4	>3.8
κ	0.30	0.35	0.40	0.45	0.50

1.6 CALCULATION OF SECTION PROPERTIES

The determination of the geometric properties of a thin gauge section can be greatly simplified if the material of the section is considered concentrated along the centre line of the sheet and the area elements replaced by straight or curved line elements. The property of the full section is obtained by multiplying that of the 'line section' by the thickness t. The method is obviously only approximate but is usually sufficiently accurate for most practical applications. The errors are small because the terms containing higher powers of t (that is t^2, t^3 etc.), involved in calculating area, modulus and inertia for example, are negligible compared to those terms containing t alone.

The properties of the two most common line elements, the straight line and the right angle bend, are given in Figure 1.22.

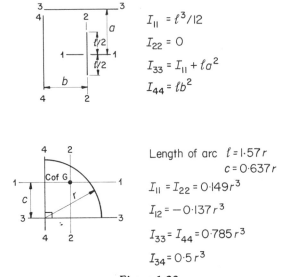

$$I_{11} = \ell^3/12$$

$$I_{22} = 0$$

$$I_{33} = I_{11} + \ell a^2$$

$$I_{44} = \ell b^2$$

Length of arc $\ell = 1.57r$

$$c = 0.637r$$

$$I_{11} = I_{22} = 0.149r^3$$

$$I_{12} = -0.137r^3$$

$$I_{33} = I_{44} = 0.785r^3$$

$$I_{34} = 0.5r^3$$

Figure 1.22

The total area A of the true section is given by $A = t\Sigma l$ and its second moment of area I (moment of inertia) is $I = I't$ where I' is the second moment of area of the line section.

In later chapters we encounter other geometric properties of the cross-section, for example the torsion constant J and the warping constant Γ. These, and other properties, can be calculated with no significant error by replacing the true cross-section (with corner radii) by a mid-plane line element (with no corner radii). Accurate values of cross-sectional properties for some common section shapes are tabulated in BS 2994.

1.7 PARTICULAR ASPECTS OF THE USE OF COLD-FORMED SECTIONS

Intrinsically, cold-formed members are thin, and this gives rise to behavioural phenomena which are not usually encountered in the more familiar hot-rolled sections used on the structural market. These phenomena obviously have to be taken into account during design. The smaller thickness of a cold-formed section can lead to problems arising from:

1 *Local buckling* — wavelike deflections in those component elements of a section which are under compression. The wavelength of these deformations is of the same order as that of the breadth of the elements. This is distinct from overall buckling (either of a strut or a beam) in which the wavelength of buckle is of the same order as the length of the member. The two modes of buckle, local and overall, can superimpose themselves in any given situation. Local buckling does not necessarily mean immediate collapse of the section, but it does reduce the stiffness of the latter causing collapse to take place at a lower load than if local buckling had not been present. The problems of local buckling of columns are discussed in Chapter 2 and of beams in Chapter 3.

2 *Low torsional stiffness* — many of the cold-formed sections that are produced have either no, or only one, axis of symmetry — Z and channels for example. This means that these sections have a natural inclination to twist under most load action. The resistance of a section to twist, as measured by its torsional stiffness J, is directly proportional to the cube of its thickness. Many light gauge cold-formed sections therefore offer only small resistance to torsional effects arising from the shape of their cross-section. Again the influence of these effects on column design is discussed in Chapter 2 and on beams in Chapter 4.

3 *Reduced bearing strengths in bolted connections* — inevitably this leads to problems in the design of many joints, and sometimes the strength of a joint may dictate the strength of a member or of a structure. The problem is becoming less crucial in many situations with the advent of improved welding techniques. However, connections are obviously a problem in the design of structures using cold-formed sections, and this is discussed more fully in Chapter 5.

4 *Corrosion* — with modern protective treatments corrosion of thin gauge sections need be no more of a problem than it is with the thicker hot-rolled sections. These treatments are listed in the British Standards Institution document DD24, different coatings being recommended depending on the severity of the corrosive atmosphere, application and maintenance of the section. For sections exposed to corrosive atmospheres, the treatments based on a protective coating of zinc followed by a recommended finishing process should give satisfactory service. Generally the strip is supplied to the cold roll formers with the protective coating already applied by the steel maker. Providing such surfaces are reasonably maintained, they should be able to provide the member with a long working life. Similarly

for structural members housed in essentially dry non-corrosive environ-ments such as the roof or floor spaces of houses, schools etc., simple treatments such as one or two coats of an epoxy based lead or oxide paint are sufficient. Obviously, before painting, the surface of the section should be properly degreased and free from rust, dirt etc. For improved adherence many paints require that the surface of the section be also primed before painting. A solution of zinc phosphate is typical of such a primer.

1.8. GENERAL COMMENTS ON DESIGN

The precise analysis of a thin walled section is to treat it as a continuous folded plate, but the mathematical complexities of such an analysis are prohibitive. Most 'exact' analyses, therefore, consider the section as being made up of an assembly of individual plates, with relevant boundary and loading conditions, such that the behaviour of the individual plates defines the behaviour of the section. The boundary conditions require compatibility of edge displacement and rotation between adjacent plates as well as equilibrium of moments and shear forces. Even the solution to this problem is difficult and neccessitates the use of a computer, so putting it beyond the means of the average design office.

For simplicity, design in most national standards is based on the assumption that at boundaries the plates support one another so that their mutual edges are maintained straight but that there are no moments or shear stresses transmitted across these boundaries. Experience has proved that this assumption gives rise to no significant error in predicting either elastic deflections or collapse loads. This assumption underlies most of the discussions that follow in this book.

The big advantage of this assumption is that the design of quite complicated shapes is reduced to a fairly simple process. This, coupled with the wide range of shapes that can be cold-formed, offers considerable flexibility in the choice and design of a section. For example, the addition of lips or swages to offset prob-lems of local buckling is discussed in Chapter 2. The design of the beam section shown in Figure 1.8 is another case in point.

There are usually two considerations in the design of a section, (1) we have one section but we require a stronger one, or, (2) what is the optimum shape from a given strip width and thickness? Comprehensive discussion of these is impossible within the limits of this book, but some of the alternatives that are available to a designer are discussed briefly below. There are basically two ways in which a section may be strengthened to carry additional load: (1) by increas-ing the strength of the steel – this has been discussed earlier in this introduction; (2) by increasing the cross-sectional area

    (a) by increasing the overall cross-section, or
    (b) by adding lips or other stiffeners, or
    (c) by increasing gauge.

Figure 1.23 shows the effect of adding an extra 25mm of strip width to a 125 x 65 x 3mm lipped channel strut. As can be seen the improvement to a 125 x 75 x 3mm lipped section (i.e. adding the material to the flanges) is much greater than that to a 150 x 65 x 3 lipped section (i.e. adding the material to the web). Also shown in Figure 1.23 is the allowable strut load curve for a 125 x 75 x 3mm unlipped channel section which has the same area (nominally) as the 125 x 65 x 3mm lipped section. As can be seen, at low strengths

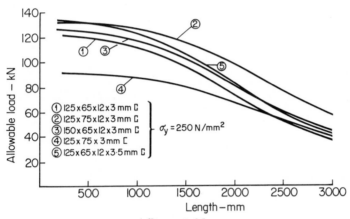

Figure 1.23

where local buckling is critical, the lipped channel is far more effective than the unlipped one, but at longer lengths the situation is reversed. The radius of gyration for the channel with the 75mm flange is larger than that for the 65mm one, and this is important in resisting primary buckling. The conclusions from this are obvious: if the strut is to be used in shorter lengths, the material is best disposed of as a lipped channel, if not, an unlipped channel is more efficient.

The same considerations apply to beams: for short spans consideration should be given to improving local buckling stiffness, but for longer spans improvement in overall stiffness will prove more beneficial.

The final alternative for increasing area is to 'up' gauge. The section with the same area as the 125 x 75 x 3 and 150 x 65 x 3mm lipped channel struts mentioned above is a 125 x 65 x 3.5 lipped channel. Safe loads for this are also shown in Figure 1.23. Leaving aside the argument about availability of this odd gauge and its consequent disproportionate cost plus the additional costs of rolling, the thicker channel is shown to be generally more efficient than the 150 x 65 but never more efficient than the 125 x 75 section.

These discussions are of necessity limited in extent, but they illustrate the freedom that is available to the designer of cold-formed structural sections.

1.9 REFERENCES

1 Bryan, E. R., *The Stressed Skin Design of Steel Buildings,* Constrado Monograph, Crosby Lockwood Staples, London, 1972.
2 Harley, W. B. and Homer, R. D., 'Methods and Economics of Fabrication in Sheet Steel', Conference on Sheet Steel in Building, RIBA, 1972.
3 Karren, K. W. and Winter, G., Proc. ASCE vol. 93, No. ST1, 1967.
4 Addendum No. 1 to BS 449 (PD 4064) 1974.
5 Uribe, J., Report No. 333, Department of Structural Engineering, Cornell University, July 1969.
6 Kaltprofile, Verlag Stahleisen MBH, Dusseldorf, 1969.

2

Design of Struts

One of the most common forms of loading is that in which the load is applied along the member. In other words, the structural component acts as a strut when the load is compressive or as a tie when the load is tensile. In this chapter we consider the behaviour of thin walled struts (see Section 4.3.5 for a discussion of the design of ties) and we consider as a preliminary example a thin walled rectangular tube loaded between the compression plattens of a testing machine.

When the tube is very short compared to its cross-sectional dimensions, that is $l \ll b_1$ or b_2 (which implies $l \doteq t$), it will have a variation of applied load against end-shortening similar to the stress/strain characteristics of a compression test specimen taken from the parent sheet material. This is shown in Figure 2.1; it is seen that up to the *crushing load* P_y there is almost linear elastic behaviour, and that with further compression the material yields and subsequently some crushing of the section in the longitudinal direction is observed. The crushing

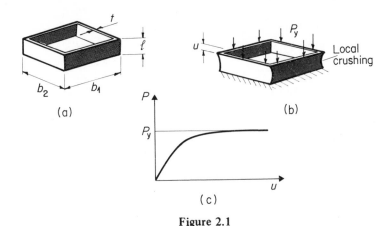

Figure 2.1

27

load in practice is usually calculated as $P_y = \sigma_y A$ where A is the cross-sectional area of the strut and σ_y is the material yield stress observed during a test on a tensile specimen of the plate material.

Now consider the corresponding behaviour of a strut whose length is of the same order as (say up to five times) the width of the widest flat side of the cross-section. As the load increases it is observed that ripples, or wavelike deflections, appear along the length of the plates forming the sides of the rectangular tube. The deformations are characterized by the fact that the corners of the tube, the junctions between the flat sides, remain essentially straight. The shape into which a short strut deforms is exemplified by the schematic diagram in Figure 2.2. The magnitude of these out-of-plane deflections increases with increasing load P in a nonlinear fashion and is accompanied by a corresponding nonlinear decrease in axial stiffness.* The load carried by the strut reaches a maximum when the corners 'crumple' or alternatively the plate elements develop a plastic mechanism, as shown in Figure 2.2(d). The strut then unloads plastically, that is it requires a decreasing load to cause increasing end-shortening. The most important feature in this behaviour is that, due to this local collapse, the strut fails at a maximum load P_{mL} which has a lower magnitude than the crushing load P_y. This type of behaviour is known as *local buckling.*

Next we consider a strut whose length is much greater than any of its cross-sectional dimensions. The deformations developed due to the application of a compressive load P are shown in Figure 2.3. In this instance the cross-section as a whole deflects into the classical column-type of deformed shape. In addition, the cross-section may be deformed into the local-wave shape discussed above. The column behaviour is generally known as *flexural buckling*, or Euler buckling, and it is important to realize that the maximum load P_{mE} that can be attained in this mode of deformation will be smaller than the local buckling maximum load P_{mL} corresponding to the same cross-section. This introduces the forms of buckling that are usually exhibited by a closed section† strut.

Open thin walled sections have a corresponding behaviour, but the load at which local buckling occurs will depend on the shape into which the section deforms. For example, a plain channel will buckle into the shape shown in Figure 2.4(a) at a lower load than that needed to cause local deflections in a corresponding lipped channel section, see Figure 2.4(b). We shall see later that this increase in local buckling load is a result of rolling a lip on to the edge of

---

* By stiffness we mean the ratio of the increment of load required to cause a corresponding unit increment of end displacement. For example, the stiffness of a linear elastic material in tension is Young's modulus E.

† A *closed section* is one in which the centre line is continuous around the cross-section, e.g. circular or rectangular sections. In contrast, an *open section* does not have a continuous centre line around the cross-section.

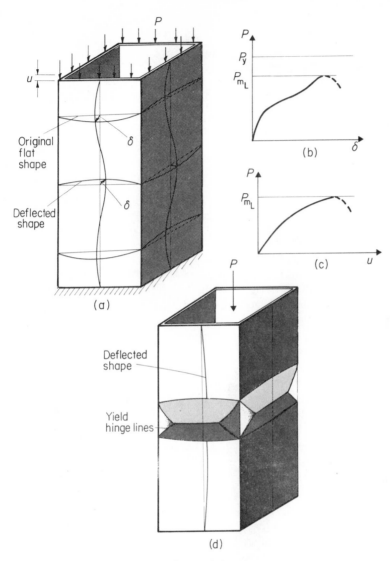

Figure 2.2

the flange thus giving it additional resistance to out-of-plane deflections.

Another form of buckling that must be considered in this discussion is that which occurs when the cross-section is weak in torsion; these are usually open sections such as angle or plain channel in which the thickness t is small. Here

COLD-FORMED SECTIONS

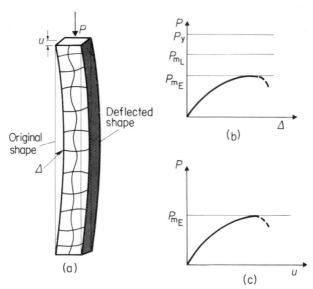

Figure 2.3

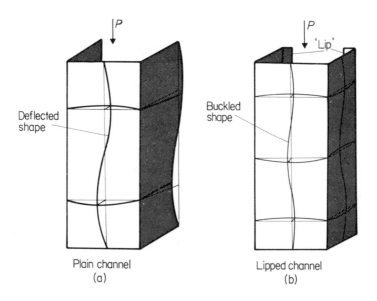

Plain channel
(a)

Lipped channel
(b)

Figure 2.4

the section may undergo purely rotational deformation about the longitudinal axis of the strut, which remains essentially straight, as shown in Figure 2.5(a) – this is known as *torsional buckling*. More commonly, where the section does not have two axes of symmetry, for example the angle section shown in Fig 2.5(b), there will be rotational and translational movement of the cross-section. This is known as *torsional – flexural* buckling.

In what follows in this chapter we shall develop a design approach that will enable us to cope with these various buckling characteristics of thin walled struts.

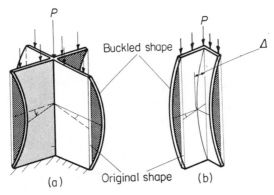

Figure 2.5

2.1 FLEXURAL BUCKLING

Before we study plate behaviour it is useful to review the more familiar analysis of a pin-ended strut. Suppose we start by considering a structural member that is perfectly straight, supported at both ends by frictionless pin joints and is subjected to a compressive load perfectly aligned with its centroidal axis. It is a well known notion that such a strut will support this load in its straight configuration up to a critical load, at which stage the strut becomes unstable and consequently deflects laterally. The *critical load*, or *Euler load*, for such a strut was derived by Euler[1], after whom this type of behaviour is often called *Euler buckling,* and can be written.

$$P_E = \frac{\pi^2 EI}{l^2} \tag{2.1}$$

where I is the second moment of area (moment of inertia) of the cross-section, l is the length of the strut between pin joints and E is Young's modulus. The corresponding critical stress is

$$\sigma_E = \frac{\pi^2 E}{(l/r)^2} \tag{2.2}$$

where r is the radius of gyration of the cross-section, $r \equiv \sqrt{(I/A)}$, and A is the cross-sectional area.

Beyond the critical load, the strut develops quite large deflections for very little increases in load (see Reference 2 for details) so that very soon the material reaches its yield stress and collapse of the strut occurs. In practical terms we can say $\sigma_m = \sigma_E$, where σ_m is the average stress on the strut at collapse. Of course if, for a strut made from a ductile material, the critical stress σ_E is greater than the yield stress σ_y, failure will occur at $\sigma_m = \sigma_y$ due to crushing. Thus we have a relationship between the collapse stress and the length of the strut given by the full line in Figure 2.6.

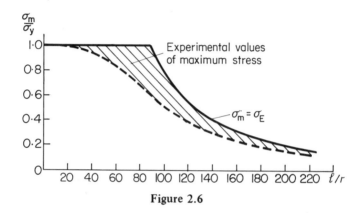

Figure 2.6

In practice, however, test results lie below this line and the lower values of these experimental collapse loads can be described by a curve such as the broken line in Figure 2.6. This reduction in the failure load is due to the presence of inevitable manufacturing defects such as initial deflections. Also, it is difficult to align the compressive load exactly with the centroidal axis, and this factor too constitutes an *imperfection*. The sum effect of the various imperfections is to cause the deflection to vary with load, as shown in Figure 2.7, and not to start at the critical load. Nevertheless, the analysis of the perfect strut does yield valuable insight[3] into the way in which the imperfect, i.e. real, strut will behave.

Taking a typical imperfect strut as one with an initial deflection of the same shape as that of the buckled perfect strut, Perry[4] derived the collapse stress as

$$\sigma_m = \frac{\sigma_y}{1 + \dfrac{ec}{r^2}\left(\dfrac{\sigma_E}{\sigma_E - \sigma_m}\right)} \tag{2.3}$$

where e is the magnitude of the initial imperfection and c is the distance of the outermost fibres (in compression) from the neutral axis. Collapse was taken to occur

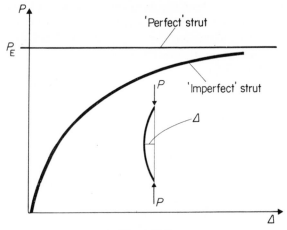

Figure 2.7

when these fibres first reached yield. Equation 2.3 is not a very useful expression for design office use because the magnitude of e will be different for every strut. However, from the analysis of a large number of tests, Robertson[5] was able to show that, by assuming the imperfection parameter ec/r^2 to vary linearly with the strut length, a curve was obtained that described adequately the lower test results. Thus, putting

$$\frac{ec}{r^2} = 0.003\frac{l}{r} \qquad (2.4)$$

into equation 2.3 gives a simple explicit relationship between σ_m and l/r (i.e. strut geometry) for a given material (defined by σ_y and E). This relationship, the *Perry-Robertson equation*, formed the basis of the column design curve in BS 449 up to the 1959 issue.

In the most recent edition of BS 449 the imperfection parameter has been modified to

$$\frac{ec}{r^2} = 0.3\left(\frac{l}{100r}\right)^2 \equiv 0.00003\frac{\pi^2 E}{\sigma_E} \qquad (2.5)$$

following recommendations made by Godfrey[6] to bring BS 449 into line with its European counterparts. Equation 2.5 was suggested by the expression

$$\frac{ec}{r^2} = 0.3\left(\frac{\sigma_y}{\sigma_E}\right) \qquad (2.6)$$

proposed by Dutheil[7] and used in the French design specification for structural steel columns.

Notice that the use of expressions such as equation 2.4 or 2.5 is in effect a process of fitting curves to test results, and as more data become available these expressions will no doubt be modified to facilitate more economic designs. Nevertheless the approach has the advantage that different types of geometric and loading imperfections for a wide range of strut length and cross-sections can all be lumped together in the same non-dimensional parameter producing a ready-to-use design method.

Of course the results quoted above are extended in practice to deal with struts having end conditions different from pin-ended simply by using the concept of *effective length* l_e. In this, the length between points of contraflexure in the strut is calculated[2] and the collapse load is then calculated based on this length. That is, in equation 2.3, etc., the actual length l is replaced by the effective length l_e.

2.2 LOCAL BUCKLING

Local buckling is so called because the length of the buckles that form are of the same order of magnitude as the cross-sectional dimensions of the structural component and are therefore usually small, or localized, when compared to the dimensions of the whole structure. For the analysis of local buckling, cold-formed sections can be considered as composed of a number of flat plates joined together at mutual edges along which the conditions of continuity and equilibrium must be satisfied. This is usually equivalent to ensuring that the moments M and M^* are in equilibrium (see Figure 2.8) and rotations ϕ and ϕ^* are equal. Before considering the buckling of complete sections it is simpler to study first the corresponding behaviour of an element plate of a section, and this we typify by a single plate supported along all four edges (see Figure 2.9) such that no deformation out-of-the-plane, i.e. in the Z direction, can occur at the edges.

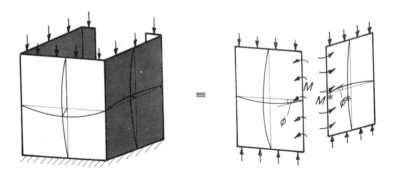

Figure 2.8

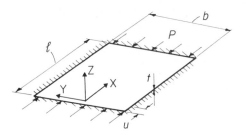

Figure 2.9

Again for simplicity, we shall consider that no rotational restraint is imposed so that the edges are in effect simply supported. We return later to the question of what effect edge restraint, supplied by contiguous plates in a thin walled section, has on the buckling load of a section.

The approach followed here for the analysis of local buckling parallels that used in Section 2.1 for the analysis of strut buckling. We first consider a perfect plate and later use these results to study the effect of imperfections. The results from tests on plates are then used to develop a simple expression that incorporates a variable imperfection parameter together with the parameters derived in the theoretical plate analysis.

2.2.1 PERFECT PLATE

Plates are ideally flat and loaded in the plane of their middle surface and, although this situation is never met in practice, it simplifies the analysis, as well as providing insight into the behaviour of real plates, to start with this *idealized*, or *perfect*, situation and then consider the effect that deviations from the perfect conditions will have on the plate behaviour.

It can be shown theoretically that if a plate is indeed flat and if the load is applied in the plane of the plate, then no out-of-plane deflections will occur until a critical level of the load is reached. At this load the flat state of the plate becomes unstable and the new stable form has deflections whose shapes are determined from the differential equation[8]

$$\frac{\partial^4 W}{\partial X^4} + 2\frac{\partial^4 W}{\partial X^2 \partial Y^2} + \frac{\partial^4 W}{\partial Y^4}$$

$$= \frac{12(1-\nu^2)}{Et^2}\left({}_0\sigma_X\,\frac{\partial^2 W}{\partial X^2} + {}_0\sigma_Y\frac{\partial^2 W}{\partial Y^2} + 2\tau_{XY}\,\frac{\partial^2 W}{\partial X\partial Y}\right)$$

where σ_X, σ_Y and τ_{XY} are the stresses existing in the flat plate due to the applied loading and ν is Poisson's ratio. The deflected shape W is determined by the physical boundary conditions of the plate, the dimensions of the plate and the loads applied to it. For the simple example considered here of a uniformly distributed load applied along two opposite edges of a rectangular simply supported plate, the buckled shape can be adequately described by

$$W = \delta \sin \frac{m\pi X}{l} \cos \frac{\pi Y}{b}, m = 1, 2, .. \quad (2.8)$$

and this is shown in Figure 2.10, where m is the number of half waves in the longitudinal direction. The values of the critical loads corresponding to these shapes are

$$P_{cr} = \frac{k\pi^2 E t^3}{12(1 - v^2)b} \quad (2.9)$$

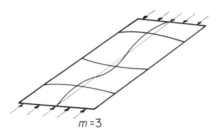

$$m = 3$$

Figure 2.10

in which k is a coefficient determined from

$$k = \left(\frac{l}{mb}\right)^2 + 2 + \left(\frac{mb}{l}\right)^2 \quad (2.10)$$

The critical stress is therefore

$$\sigma_{cr} = \frac{k\pi^2 E}{12(1 - v^2)(b/t)^2} \quad (2.11)$$

where b/t is known as the *thinness ratio* of the plate.

The form of equation 2.11 compares closely with that for the corresponding strut expression, equation 2.2, the thinness ratio of the plate corresponding to the slenderness ratio of the strut.

The coefficient k is shown graphically in Figure 2.11 where we see that for a plate with $l/b > 4$ the buckles will approximate to square wave forms (that is the number of half waves m very nearly equals the aspect ratio l/b), so that $k \simeq 4$ and the critical load will be

$$P_{cr} = \frac{\pi^2 E t^3}{3(1 - v^2)b} \quad (2.12)$$

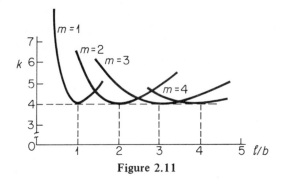

Figure 2.11

Notice that this load is a linear function of the stiffness of the material, that is Young's modulus, and is not influenced by the strength of the material, that is its yield stress.

The above analysis has been concerned with *elastic buckling* and implies that the critical load will be smaller than the squash load,

$$P_{cr}\left(\equiv \frac{\pi^2 E t^3}{3(1-\nu^2)b}\right) < P_y \ (\equiv \sigma_y bt)$$

This implies in turn that the geometry of the plate must be such that

$$\frac{b}{t} > \left(\frac{\pi^2}{3(1-\nu^2)} \frac{E}{\sigma_y}\right)^{\frac{1}{2}} \tag{2.13}$$

For $E = 200\,\text{kN/mm}^2$, $\sigma_y = 250\,\text{N/mm}^2$, $\nu = 0.3$, we must have $b/t > 54$ for elastic buckling to take place. Other conditions of loading and boundary conditions have corresponding elastic buckling limits, and these, together with critical loads and buckling deflected shapes, have been collated in Reference 8.

It can be shown theoretically that, provided $P_{cr} < P_y$, a plate can support loads in excess of the critical load, the thinner the plate for a given width (that is the greater the b/t ratio) the greater will be the excess. In very many cases the small magnitudes of out-of-plane deflections that occur in the post-buckling range of loading are not significant in practice and it would be uneconomical not to take this reserve of strength into account in using such structural elements. Notice that this is in contrast to a strut which, even in its perfect state, can support virtually no increase of load above the critical load.

A very simple model to explain the mechanism whereby the plate can support a load even though the plate has deflected is the grid of longitudinal and transverse bars shown in Figure 2.12 due to Winter.[9]

It was shown earlier that a rectangular perfect plate tends to buckle into nearly square waves, and this is confirmed by tests on real plates. By taking a

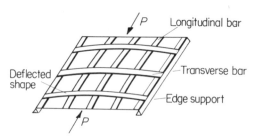

Figure 2.12

square plate and studying its behaviour, by means of the model, we are in effect considering a typical portion of a rectangular plate, and the results we obtain are therefore generally applicable to a wide range of plate geometries.

Suppose the vertical bars in the model, which represent strips of the plate, are straight and are loaded along their axes, they will act as perfect struts and be able to support increasing compressive loading without developing any lateral (out-of-plane) deflections, but eventually they will reach a critical value of the load (equation 2.2) and simultaneously begin to buckle. Now, if the vertical bars were not attached to the horizontal bars, they would in effect be simply supported struts and, as the analysis in Section 2.1 showed, they would develop very large deflections and no further loading could be sustained. But, of course, as the vertical bars deflect, so also must the horizontal bars, and in so doing they must stretch. Thus the horizontal bars by the action of stretching tend to restrain the buckling deflections of the vertical struts. The model, or the plate which it represents, will therefore not collapse when its critical load is reached; in contrast to a strut, a plate will merely develop slight buckling deflections at the critical load and will continue to carry increasing load. The theoretical variation of the maximum deflection δ (at the centre of the plate) with increasing load P is shown in Figure 2.13(a); these results were obtained from an approximate analysis.[10]

While the bars are still straight ($P < P_{cr}$) each will carry a proportionate amount $P/5$ of the applied load. But when the load exceeds the critical load the bars buckle and each deflects a different amount, the ones nearest the centre of the model deflecting the most. This means that each bar will now carry a different proportion of the load; the bars at the edges of the model which are supported by the edge conditions remain almost straight and thus carry more of the increasing load than the middle bars which 'get away from the load'.[9] The tension and compression loads in the bars of the model are analogous to the membrane stresses in the middle surface of the plate. These latter are primarily compressive down the length of the plate and tensile across the width. Due to the development of out-of-plane deflections the compressive membrane stresses will redis-

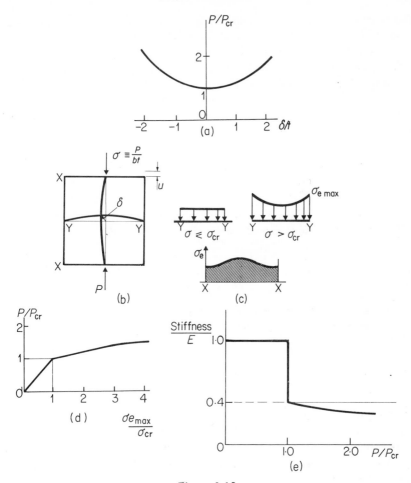

Figure 2.13

tribute and become concentrated at the supported unloaded edge of the plate, as shown in Figure 2.13(c).

Analysis of the post-buckled plate is a complicated process and no exact closed form results have been obtained for compressed plates. However, an approximate expression that describes the variation of the deflection at the centre of the plate with applied load (for loads greater than the critical load) is[11]

$$\frac{\delta}{t} = 2.1 \left(\frac{P}{P_{cr}} - 1 \right)^{\frac{1}{2}} + 0.02 \left(\frac{P}{P_{cr}} - 1 \right)^{\frac{3}{2}} \qquad (2.14)$$

This equation is sufficiently accurate for engineering purposes for loads up to about four times the critical load, that is $1 \leqslant P/P_{cr} \lesssim 4$.

It was mentioned above that the stresses in the plate redistribute as the plate buckles; the variation of the longitudinal direct stress σ_e along the unloaded edge is shown in Figure 2.13(c). Thus the maximum direct stress σ_{emax} occurs midway down the unloaded edge. Once more because of the mathematical complexity of the problem, no exact expression has been derived relating the applied load to the maximum edge stress, but an approximate solution[12] is

$$\frac{\sigma_{emax}}{\sigma_{cr}} = \frac{P}{P_{cr}} + \left[2.83 \left(\frac{P}{P_{cr}} - 1 \right) + 0.52 \left(\frac{P}{P_{cr}} - 1 \right)^2 \right]_{(P>P_{cr})} \qquad (2.15)$$

This is shown graphically in Figure 2.13(d). The amount u by which the plate will shorten as the load is applied is approximately given by[11]

$$\frac{u}{u_{cr}} = \frac{P}{P_{cr}} + \left[1.45 \left(\frac{P}{P_{cr}} - 1 \right) + 0.38 \left(\frac{P}{P_{cr}} - 1 \right)^2 \right]_{(P>P_{cr})}$$

where $u_{cr} \equiv P_{cr}/Et$. An interesting result of this equation arises when we consider the stiffness S of the plate; the *rate* of unit end-shortening u/b with applied average stress σ_{av}, that is

$$S = \frac{d\sigma_{av}}{d(u/b)}$$

For the initially flat plate, $\sigma < \sigma_{cr}$, we have $S = E$ (Young's modulus for the material) but as soon as the plate begins to buckle, $\sigma = \sigma_{cr}$, the stiffness of the plate decreases to about 40% of this value, that is $S = 0.4E$ (see Figure 2.13(e)). This factor is of considerable importance if the plate is a member in a redundant structure in which the distribution of internal forces depends on the stiffness of the various members. The corresponding post-buckling stiffness of a strut is, for all practical purposes, equal to zero.

Of course the basic assumption in the analysis, from which the above results are quoted, is that the material is linear elastic. But with increasing loads the stresses increase rapidly and eventually reach the value of the yield stress for the plate material, and plastic deformations will take place. The analysis of a plate which incorporates large deflections and plasticity is very difficult indeed and is possible for particular cases only with the aid of computers. An important result of one such study[13] was that the occurrence of the maximum load closely coincided with the load at which the maximum edge stress reached the yield stress. For a ductile material with extensive plastic deformation at the yield stress (as is common in cold-rolled sections), the load which the plate can carry diminishes after the maximum load is reached. Now, from equation 2.15 and with $P = P_{mL}$ when $\sigma_{emax} = \sigma_y$ we have

$$\frac{P_{mL}}{P_y} = 0.36 + 0.83\left(\frac{P_{cr}}{P_y}\right) - 0.19\frac{(P_{mL}/P_y)^2}{(P_{cr}/P_y)} \qquad (2.16)$$

which is shown in Figure 2.14. Equation 2.16 is of course only valid for $P_y > P_{cr}$ since for $P_y < P_{cr}$ we have $P_{mL} = P_y$, that is failure by crushing. The ratio P_{mL}/P_y is a measure of the reduction of the maximum load P_{mL} (from the squash load P_y) caused by the onset of buckling. The parameter $(P_y/P_{cr})^{1/2}$ incorporates the material properties, plate geometry and loading, and for a given material and

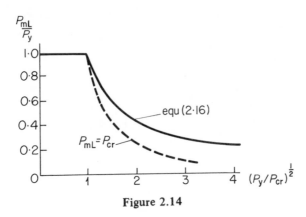

Figure 2.14

plate loading is proportional to the plate width to thickness ratio. For a square simply supported plate

$$\left(\frac{P_y}{P_{cr}}\right)^{\frac{1}{2}} \simeq 0.019\left(\frac{b}{t}\right) \quad \text{(for } E = 200\,\text{kN/mm}^2\,;\, \sigma_y = 250\,\text{N/mm}^2\text{)}$$

Notice the similarity in form of Figure 2.14 for a plate and Figure 2.6 for a strut. The important difference is that the strut can never carry a load greater than P_E, but a thin plate has a reserve of strength beyond the critical load.

2.2.2 EFFECT OF IMPERFECTIONS

When we carry out a test on a plate we observe that the magnitude of the out-of-plane deflections grows from the beginning of load application. In other words there is no load that we can identify as the critical load. This lack of agreement between the observed behaviour of a real plate and that predicted on the basis of a perfect flat plate is of course due to the fact that real plates are not flat prior to loading. Owing to the process of manufacture some initial deflections will exist, these will be different in magnitude and shape from one plate to another. In order to analyse the plate precisely we need to be able to specify the initial deflections but we do not know these until the plate exists. Therefore, to simplify the analysis, the initial deflections are chosen to coincide in shape with the deflected shape that would occur in a buckled perfect plate. Hence the long rectangular plate in question can be regarded as a square panel with an initial deflection of W_0

$$W_0 = \epsilon \sin \frac{\pi X}{b} \cos \frac{\pi Y}{b} \tag{2.17}$$

where ϵ is the amplitude of the initial imperfection.

Analysis[12] shows that the magnitude δ of the buckles can now be determined from

$$\left[\left(\frac{\delta}{t} \right)^2 - \left(\frac{\epsilon}{t} \right)^2 \right]^{\frac{1}{2}} = 2.1 \left(\frac{P}{P_{cr}} - 1 + \frac{\epsilon}{\delta} \right)^{\frac{1}{2}} + 0.02 \left(\frac{P}{P_{cr}} - 1 + \frac{\epsilon}{\delta} \right)^{\frac{3}{2}} \tag{2.18}$$

and that for a given buckle amplitude the end-shortening is

$$\frac{u}{u_{cr}} = \frac{P}{P_{cr}} + 1.45 \left(\frac{P}{P_{cr}} - 1 + \frac{\epsilon}{\delta} \right) + 0.38 \left(\frac{P}{P_{cr}} - 1 + \frac{\epsilon}{\delta} \right)^2 \tag{2.19}$$

Notice the similarity of form of these equations with equations 2.14 and 2.15. Equations 2.18 and 2.19 are shown in Figure 2.15; the significant feature is that at high values of P/P_{cr} the effect of initial deflection diminishes, and also that, although there is now no practical load we can identify as the critical load, nevertheless this parameter P_{cr} still has an important place in the description of the plate behaviour. The value of the maximum edge stress, which still occurs mid-way along the unloaded edges, is now

$$\frac{\sigma_{e max}}{\sigma_{cr}} = \frac{P}{P_{cr}} + 2.83 \left(\frac{P}{P_{cr}} - 1 + \frac{\epsilon}{\delta} \right) + 0.52 \left(\frac{P}{P_{cr}} - 1 + \frac{\epsilon}{\delta} \right)^2 \tag{2.20}$$

If once again we assume that failure occurs when the stress reaches the yield value, that is at $\sigma_{e max} = \sigma_y$, and if we take the value of the deflection at this load to be δ_m, the maximum load can be calculated using

$$\left(\frac{\delta_m}{t} \right)^3 - \frac{\delta_m}{t} \left[\left(\frac{\epsilon}{t} \right)^2 + 4.25 \left(\frac{P_{mL}}{P_{cr}} - 1 \right) \right] - 4.25 \left(\frac{\epsilon}{t} \right) = 0 \tag{2.21a}$$

$$\frac{P_y}{P_{cr}} = \frac{P_{mL}}{P_{cr}} + 2.83 \left(\frac{P_{mL}}{P_{cr}} - 1 + \frac{\epsilon}{\delta_m} \right) + 0.52 \left(\frac{P_{mL}}{P_{cr}} - 1 + \frac{\epsilon}{\delta_m} \right)^2 \tag{2.21b}$$

Figure 2.16(a) shows the curves of collapse load for specified values of initial deflection ϵ. But values of initial deflection will vary according to the plate dimension,

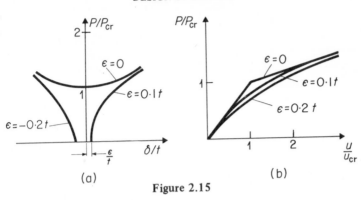

Figure 2.15

for example a thick plate will have a smaller initial deflection relative to the thickness than a thinner plate having the same width. Thus, much in the same manner as Robertson for struts, we can arrange for ϵ/t to be a parameter of the plate geometry. A suitable description is $\epsilon/t = \beta(P_y/P_{cr})$, in which β is a constant that can be adjusted to fit experimental results. Figure 2.16(b) shows that $\beta = 0.2$ will provide a good description of the lower values from a number of tests on single plates and square tubes. These latter structural elements because of their symmetry have no constraining moment from one component to another and therefore correspond closely to the individual simply supported plates analysed here.

2.2.3 EFFECTIVE WIDTH
The load reduction P_{mL}/P_y is equivalent to the local buckling stress factor C_L used in the British Standards Specification BS 449. However, there is another basic approach to assessing the reduction of efficiency of a compressed plate. This is to assume that the maximum edge stress acts over a certain portion of the plate width. The portion over which this stress acts is called the 'effective width' b_e of the plate.

Figure 2.17 shows diagrammatically the concepts of effective width and stress reduction. The actual stress pattern is shown by the full curved line and the average applied stress σ_{av} is shown by the horizontal dotted line. The total load P on the plate is given by the area beneath the stress curves. The value of σ_{av} is therefore such that $\sigma_{av} \times b \times t$ is equal to the area under the actual stress pattern. The effective width of the plate b_e is found by evaluating the width of the plate required to carry the total load if σ_e acts over this width, therefore $b_e \times \sigma_e \times t$ is equal to the area under the actual stress pattern.

Thus

$$P = \sigma_{av} \times b \times t \quad \text{and} \quad P = \sigma_e \times b_e \times t$$

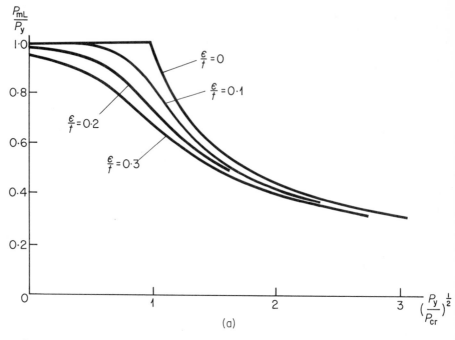

(a)

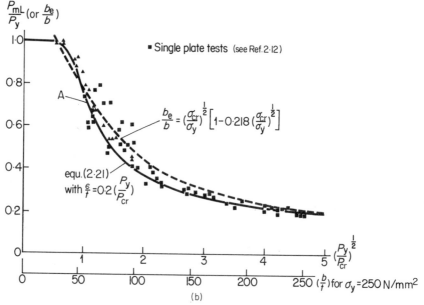

(b)

Figure 2.16

which gives $\qquad \sigma_e \times b_e = \sigma_{av} \times b, \quad$ or $\quad \dfrac{b_e}{b} = \dfrac{\sigma_{av}}{\sigma_e}$

At collapse $\qquad \dfrac{b_e}{b} = \dfrac{\sigma_{av}}{\sigma_y} = \dfrac{P_{mL}}{P_y} = C_L$

Thus at the collapse condition we have that the local buckling stress factor C_L can be interpreted as the ratio of an effective width to the full width of the plate. The current American practice[9] is to use the effective width formulation

$$\frac{b_e}{t} = 1.9 \left[\frac{E}{\sigma_e}\right]^{\frac{1}{2}} \left[1 - \frac{0.415}{b/t}\left(\frac{E}{\sigma_e}\right)^{\frac{1}{2}}\right] \qquad (2.22)$$

This was derived on an empirical basis from a number of tests, and with the maximum edge stress equal to the yield stress gives

$$\frac{b_e}{b} \equiv \frac{P_{mL}}{P_y} = \left(\frac{P_{cr}}{P_y}\right)^{\frac{1}{2}} \left[1 - 0.218\left(\frac{P_{cr}}{P_y}\right)^{\frac{1}{2}}\right]$$

But equation 2.22 can also be used for loads less than the maximum load provided the appropriate value of σ_e is known.

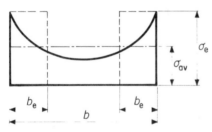

Figure 2.17

From a practical viewpoint there is little to choose between the two concepts discussed in this section; either a reduced stress (given by $C_L \times \sigma_y$) acts across the full width of the plate, or the maximum stress acts across part b_e of the plate only. However, it may be noticed that the effective width concept apparently gives a more true to life picture of the actual behaviour of a compressed plate, since it can be seen from Figure 2.17 that the actual stress pattern indeed shows the stress level over part of the plate width to be low and therefore that part is behaving less efficiently than the rest of the plate width.

2.2.4 PLATES WITH A FREE EDGE
So far we have considered plates that are supported along both unloaded edges so that no out-of-plane deflections can occur. But some cold-formed sections

may have component plates which have no such edge support. In other words they may have one edge that is free from any lateral or torsional restraint. Such a plate, were it perfectly flat, could support increasing load up to its critical load at which stage it must begin to deflect. The theoretical value of the critical load is given by equation 2.9 but now the value of the coefficient k will be different from that in equation 2.10. For a plate whose supported unloaded edge is simply supported the coefficient is given approximately as

$$k = 0.45 + 0.98 \left(\frac{b}{l}\right)^2$$

This is shown graphically as curve 1 in Figure 2.18(a), and curve 2 shows the corresponding values for a plate in which the supported edge is clamped against any rotation.

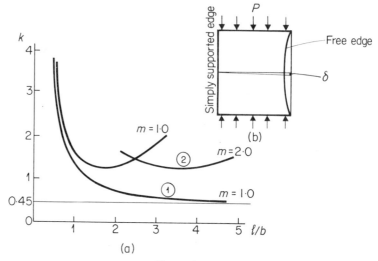

Figure 2.18

As can be seen from the figure, a long plate buckles into several half waves if the supported edge is clamped, and into a single half wave if it is simply supported. This latter plate has a buckling coefficient which approaches the value of 0.45 when the plate is very long. In practice, the support condition on the supported edge will lie between the clamped and simple support conditions, and it therefore ensures safe design for any section containing unstiffened plates (that is plates with one free edge) if the minimum value of k is used, $k = 0.45$. The deflected shape of a simply supported free plate is shown in Figure 2.18(b).

A large deflection analysis has been carried out[14] to determine the maximum load P_{mL} such a plate could support and it was shown that the parameters

(P_{mL}/P_y) and (P_y/P_{cr}) are relevant to this form of boundary condition. Tests on cruciform sections and angle sections, which both should provide simple support along the supported edge, show that, although equation 2.21 with

$$\frac{\epsilon}{t} = 0.2 \frac{P_y}{P_{cr}}$$

was not derived for a plate with a free edge, it nevertheless does provide a curve that conservatively describes the lower values of the test results (see Figure 2.19).

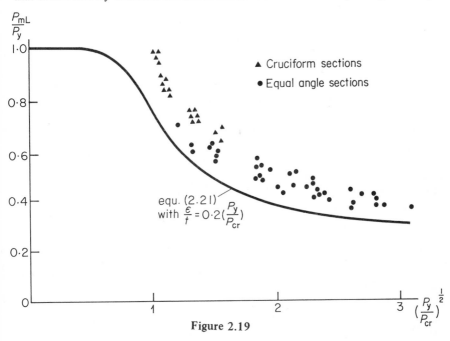

Figure 2.19

2.2.5 EFFECTS OF LIPS AND SWAGEING ON PLATE BUCKLING
Lips may be used to increase the buckling strength of compression elements which would otherwise have free edges. The purpose of a lip in this respect is to hold the edge straight when the element has buckled locally, thus imposing a condition akin to simple support on this edge. To achieve this purpose, the lip itself must be of sufficient width[15] to avoid the possibility of out-of-plane waving as shown in Figure 2.20. It has been observed in practice that, in order to prevent such deflections occurring, the lip must have a width b_e of at least $b_e \leqslant b/5$, where b is the width of the adjacent compression element.

The buckling strength of a section will increase as the lip width increases, and strictly the lip should be treated in the buckling analysis as an unstiffened plate element. But such elements are fully effective only for width to thickness ratios

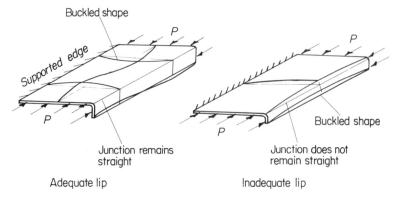

Figure 2.20

less than 10. Hence for economic design, as well as simplifying the analysis, the lip is usually dimensioned such that $b_e/t \leqslant 10$.

The maximum average stress of a stiffened plate increases as the critical stress increases (see Figure 2.16), and the critical stress varies inversely as the square of the plate width. It follows therefore that a narrow plate will be more fully effective than a wide plate having the same thickness. In practice it may often be valuable to be able to subdivide a wide plate into a number of narrower strips and thus increase its effectiveness. This can be achieved in fabricated plate construction by welding on longitudinal stiffeners and in cold forming by rolling, or swageing, ribs into the cross-section, see Figure 2.21 for example. This has the effect of replacing a plate of width b by plates of width b^*. But it considerably increases the complexity of the analysis of the buckling behaviour and usually the dimensions of the ribs are determined from manufacturers' tests. However, from the above discussion on lip widths it follows that as a preliminary guide the depth of the rib should be about $b^*/5$.

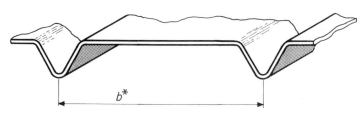

Figure 2.21

2.2.6 BUCKLING OF SECTIONS

The first characteristic we must consider is the critical load since from our study on single plates it is evident that this governs the behaviour of the buckled plate. For a section the critical load is a function of the geometry of the complete section and it can be derived from a knowledge of the critical loads of the component plates having rotational restraint along the edges[16,17] and by matching the restraints at the mutual corners. Alternatively it can be determined by considering the section as a whole and calculating the critical load by the same method as for a single plate but with much more complicated arithmetic.[18-21] Bulson[8] has collated the critical loads for a wide variety of structural shapes. Figure 2.22 shows values of the coefficient k for the more common cold-formed structural shapes which can be used to calculate the relevant critical load

$$P_{cr} = \frac{k\pi^2 EA}{12(1-\nu^2)(b_w/t)^2} \qquad (2.23)$$

where b_w is the width of one of the component plates (see Figure 2.22) and A is the cross-sectional area of the cold-formed section.

If the section were formed of perfect plates it would begin to deflect at the critical load and the interaction between the plates would adjust to maintain

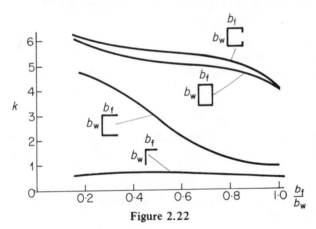

Figure 2.22

equilibrium and compatibility at the corners as the deflections increased. From our study of individual plates it would seem probable that the section would continue to carry increasing loading until the stress at the corner, which would be the maximum value in the section, reached yield. Thus we would expect the parameters P_{mL}/P_y and P_y/P_{cr} again to govern the collapse behaviour. This was argued intuitively by Chilver[22]. But because of the variation of edge restraint for various geometries and possibly due to different imperfections, a suitable imperfection magnitude for use with equation 2.21 is now $\epsilon/t = 0.3(P_y/P_{cr})^{\frac{1}{2}}$. As

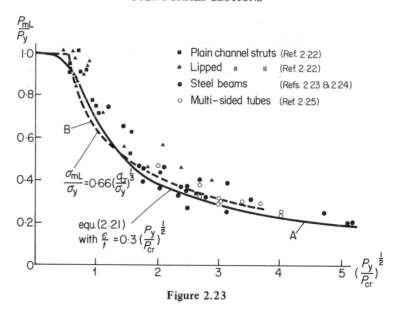

Figure 2.23

shown in Figure 2.23, this curve A provides a good description of the lower test results of Chilver[22], Winter[23,24] and Bulson.[25]

On a purely empirical basis Chilver[22] derived the relationship

$$\frac{\sigma_{mL}}{\sigma_y} = 0.66 \left(\frac{\sigma_{cr}}{\sigma_y}\right)^{\frac{1}{3}}$$

to describe the lower values of test results in the range

$$0.6 < \left(\frac{\sigma_y}{\sigma_{cr}}\right)^{\frac{1}{2}} < 2.8,$$

this is shown as curve B in Figure 2.23. Hence, the maximum average stress can be calculated using

$$\sigma_{mL} = 0.66 \, \sigma_y^{\frac{2}{3}} \left[\frac{\pi^2 E}{12 \, (1-\nu^2)}\right]^{\frac{1}{3}} k^{\frac{1}{3}} \left[\frac{t}{b_w}\right]^{\frac{2}{3}}, \tag{2.24}$$

where k is evaluated in Figure 2.22

The experimental study (reported in Reference 25) was particularly interesting since it consisted of compressing to failure a number of tubes having the same perimeter. The perimeter was formed into a number of flat sides and it was shown that tubes having between four and 18 sides buckled as if they were composed of a number of flat plates. For a greater number of sides, however,

the longitudinal corners did not remain straight and a type of buckling similar
to that for a circular cylindrical shell occurred. This failure load was less than
that predicted for a flat sided tube; thus it may be inferred that provided the
angle formed at the junction of two plate elements is less than 135° (measured
at the inside radius) the section will behave as a collection of plates.

To simplify the design procedure in situations where the critical load is not
known, for a complex shape for example, an approximate collapse load for the
section can be synthesized from the collapse loads of the element plates on the
basis that they are either simply supported along both edges or simply supported
along one edge and free along the other.

For a plate of width b to be considered as simply supported, that is adequately
restrained against lateral deflection, the adjacent plate should have a width of at
least $b/5$. If the adjacent plate has a width less than $b/5$ the edge must be
considered a free edge.

Example 2.1
As an example of the determination of the maximum load, we consider the
lipped channel shown in Figure 2.24 made from 3 mm thick steel with

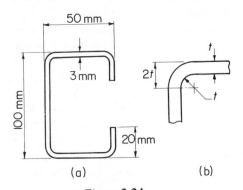

(a)   (b)

Figure 2.24

σ_y = 350 N/mm². First using the more accurate method; the centre line
dimensions give

$$\frac{b_f}{b_w} = \frac{50 - 3}{100 - 3} = 0.48$$

and hence from Figure 2.22 k = 5.5. The critical stress is therefore

$$\sigma_{cr} = \frac{k\pi^2 E}{12(1 - \nu^2)}\left(\frac{t}{b_w}\right)^2 = \frac{5.5 \times \pi^2 \times 200 \times 10^3}{12 \times 0.91}\left(\frac{1}{32.3}\right)^2 = 951 \text{ N/mm}^2.$$

This gives $(\sigma_y/\sigma_{cr})^{\frac{1}{2}} = 0.61$ and hence, from Figure 2.23 (curve A), we have $P_{mL}/P_y = 0.92$. The maximum average stress is therefore

$$\sigma_{mL} = 0.92 \times 350 = 320 \text{ N/mm}^2$$

Now for comparison we calculate the maximum stress by synthesizing the effectiveness of the component plates. The width of a plate is taken as the portion between the tangent points of the corner radii and, since the inside radius is usually equal to the plate thickness, this is about $2t$ from the outer surface of the section (see Figure 2.24(b)). The flat-width to thickness ratio b/t is then calculated and hence the ratio $(\sigma_y/\sigma_{cr})^{\frac{1}{2}}$ can be obtained from

$$\left(\frac{\sigma_y}{\sigma_{cr}}\right)^{\frac{1}{2}} = 1.19 \times 10^{-3} \, \sigma_y^{\frac{1}{2}} \left(\frac{b}{t}\right), \qquad (2.25a)$$

for a plate supported on both edges, or

$$\left(\frac{\sigma_y}{\sigma_{cr}}\right)^{\frac{1}{2}} = 3.52 \times 10^{-3} \, \sigma_y^{\frac{1}{2}} \left(\frac{b}{t}\right) \qquad (2.25b)$$

for a plate supported along only one edge. Having calculated the values of these parameters for each component plate the corresponding values of maximum loads can be obtained from curve A in Figure 2.16(b). The maximum load that can be carried by the section is taken to be the sum of the maximum loads of the individual component plates.

A slightly different, but essentially equivalent, procedure is used in BS 449 Addendum 1. In Table 5 of that document we have values of $C_L(\equiv P_{mL}/P_y)$ tabulated for a wide range of plate thinness ratios b/t. The table has been compiled using the same formula as that described graphically by curve A in Figure 2.16(b) and the b/t values relate to a yield stress $\sigma_y = 250$ N/mm². If, as in this example, we have a material with a different yield stress an equivalent thinness ratio $(b/t)_{equiv}$ must be obtained. From equation 2.25 we see that to give the correct value of $(\sigma_y/\sigma_{cr})^{\frac{1}{2}}$ the equivalent thinness ratio in this example should be calculated as

$$\left(\frac{b}{t}\right)_{equiv} = \frac{b}{t} \sqrt{\left(\frac{350}{250}\right)}$$

Hence from BS 449 Addendum 1 we can obtain the correct value of C_L for each component plate. If now we sum the effective widths $C_L b$ of all such plates and divide by the sum of the full plate widths, we obtain a value for the mean stress reduction factor C_{Lm}. That is

$$C_{Lm} = \frac{\Sigma C_L b}{\Sigma b}$$

This procedure is shown in detail in Table 2.1

Table 2.1

Plate component	Condition	b	b/t	$(b/t)_{equiv}$	C_L	$C_L b$
1	unstiffened	14	4.7	5.6	1.0	14
2	stiffened	38	12.7	15.0	1.0	38
3	stiffened	88	29.3	34.7	0.98	86.2
4	stiffened	38	12.7	15.0	1.0	38
5	unstiffened	14	4.7	5.6	1.0	14
		$\Sigma = 192$				$\Sigma = 190.2$

Mean effectiveness of cross-section $C_{Lm} = 0.99$,

$$\sigma_{mL} = \sigma_y \times C_{Lm} = 347 \text{ N/mm}^2$$

The approximate method thus gives a higher value than that calculated from the lower bound of the test results in Figure 2.23. It has been shown more generally by Reiss and Chilver[26] that for various geometries and plate thicknesses the approximate method tends to overestimate the maximum load when compared to the more conservative estimate based on equation 2.24. For example, for the section with the same profile as in Figure 2.24 but having a thickness of 1.5mm, the effective width method gives a value for the section mean effectiveness of $C_{Lm} = 0.72$ while using Figure 2.23 (curve A) we have $P_{mL}/P_y (\equiv C_{Lm}) = 0.59$.

2.3 OVERALL BUCKLING

We will now return to a consideration of the type of buckling that is common to cold-rolled and hot-rolled structural sections alike — that is *overall buckling*, or *Euler buckling* as it is sometimes called. But because of the large variety of section shapes that can be manufactured by rolling or pressing it is necessary to treat a more general form of buckling than that in Section 2.1. This involves considering the situation where no axis of symmetry exists, and the section twists as well as bowing sideways and we have *torsional–flexural* buckling. This behaviour is less prevalent in hot-rolled sections because of their higher torsional rigidity and also because they usually have at least one axis of symmetry.

A very full presentation of the theory for torsional–flexural buckling is presented by Timoshenko,[27,28] and it is sufficient to present here only the results of the theory. If we consider some general open thin walled cross-section, see Figure. 2.25, subject to a longitudinal load applied along its centroidal axis O, and by assuming that during buckling the cross-section will not distort but will translate and rotate, it is possible to derive a differential equation describing the equilibrium of the strut. Now if it is also assumed that the strut

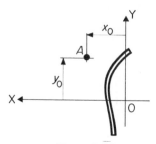

Figure 2.25

deflections are of simple sinusoidal shape, the critical loads P_{cr} for the various forms of buckling may be determined from the cubic equation.

$$\frac{I_c}{A}(P_{cr} - P_{EY})(P_{cr} - P_{EX})(P_{cr} - P_\phi)$$
$$- P_{cr}^2\,[y_0^2\,(P_{cr} - P_{EX}) + x_0^2\,(P_{cr} - P_{EY})] = 0 \qquad (2.26)$$

where XX and YY are the principal centroidal axes (that is I_X and I_Y are respectively the maximum and minimum values of the second moments of area) and x_0, y_0 are the distance along the ordinates X and Y of the shear centre A from the centroid; I_c is the polar second moment of area about the shear centre c; $I_c = I_x + I_y + A(x_0^2 + y_0^2)$ (see Chapter 4); P_{EX} ($\equiv \pi^2\,EI_X/l^2$) is the critical load for flexural buckling about the X axis; P_{EY} ($\equiv \pi^2\,EI_Y/l^2$) is the critical load for flexural buckling about the Y axis, and P_ϕ ($\equiv A(J + \Gamma\pi^2/l^2)/I_0$) is the critical load for torsional buckling.

The coefficients J and Γ are functions of the cross-sectional dimensions of the strut; J is the torsional rigidity and Γ is the warping rigidity of the strut (see Chapter 4).

For any section with no axis of symmetry it is first necessary to calculate the values of the coefficients and then to solve equation 2.26 to determine the smallest value of P_{cr} at which buckling will occur. It turns out that the lowest critical load is always smaller, and the highest root of equation 2.26 is always greater, than P_ϕ, P_{EX} or P_{EY} and the third root is intermediate between P_ϕ and P_{EX} or P_{EY}. Hence by taking the possibility of twisting deformation into account we find that the actual critical load is smaller than either of the Euler loads or the purely torsional critical load.

2.3.1 SECTION WITH ONE AXIS OF SYMMETRY
If we have a section with two axes of symmetry, such as an I section, the shear centre coincides with the centroid and x_0 and y_0 are each zero. Hence equation

2.26 becomes

$$(P_{cr} - P_{EX})(P_{cr} - P_{EY})(P_{cr} - P_\phi) = 0 \qquad (2.27)$$

and the various modes of buckling occur independently of one another. The lowest critical load therefore coincides with the lowest of the loads P_{EX}, P_{EY}, P_ϕ.

If we have a section with one axis of symmetry we have one form of buckling that will be in a purely flexural mode while the remaining two will comprise flexural and torsional deformations. As an example, we consider an equal angle column[29], Figure 2.26, with the principal centroidal axes XX and YY. The axis XX is an axis of symmetry so that buckling can occur by bending about YY without any rotational distortion. The Euler load for this mode of deformation is

$$P_{EY} = \frac{\pi^2 \, EI}{l^2} ; I_Y = \frac{1}{12} b^3 t$$

$$P_{EY} = \frac{\pi^2 \, Eb^3 t}{12 \, l^2} \qquad (2.28)$$

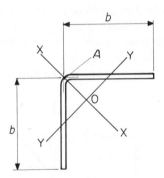

Figure 2.26

$$I_c = I_X + I_Y + A x_0^2 \simeq \frac{1}{3} b^3 t + \frac{1}{12} b^3 t + \frac{1}{4} b^3 t = \frac{2}{3} b^3 t$$

$$I_0 \equiv I_X + I_Y = \frac{5}{12} b^3 t$$

$$\frac{I_c}{I_0} = 1.6$$

For the remaining two buckling modes we have

$$\frac{I_0}{I_c} P_{cr}^2 - (P_{EX} + P_\phi) P_{cr} + P_{EX} P_\phi = 0 \qquad (2.29)$$

where

$$P_{EX} = \frac{\pi^2 E I_X}{l^2} = \frac{\pi^2 E b^3 t}{3 l^2} \qquad (2.30a)$$

For no warping restraint,

$$P_\phi = \frac{A}{I_c} J = \frac{2 G t^3}{b} \qquad (2.30b)$$

Torsional–flexural buckling will only occur when the smaller root of equation 2.29 has a value less than the magnitude of the smaller Euler load, that is $P_{EY} = \pi^2 E b^3 t / 12 \, l^2$. This occurs when

$$\frac{l}{r_Y} \leqslant \frac{3\pi}{2\sqrt{2}} \sqrt{\left(\frac{E}{G}\right)} \frac{b}{t} \qquad (2.31)$$

where r_Y is the radius of gyration defined by $r_Y \equiv \sqrt{(I_Y/A)}$. Taking $E = 2.6 \, G$ we find that if $(l/r_Y)(t/b) < 5.38$ then torsional flexural buckling will occur, otherwise purely flexural buckling will take place. This is shown diagrammatically in Figure 2.27;[29] also indicated is the region in which the strut will reach the crushing load before any of the critical loads.

It is convenient for presentation to modify the Euler load for a column which

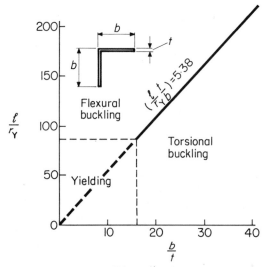

Figure 2.27

is liable to torsional–flexural buckling, rather than to introduce a new formulation. This we can do by introducing a coefficient α into the Euler load formula the procedure being exemplified by the following case. Suppose we have an equal angle strut whose geometry is

$$\frac{l}{r_Y} \frac{t}{b} = 4$$

From equation 2.30 this implies that $P_\phi/P_{EX} = 0.16$ and from equation 2.29 the lower root has a value of $P_{cr}/P_{EX} = 0.15$, that is

$$P_{cr} = 0.15 \frac{\pi^2 EIx}{l^2} = 0.6 \frac{\pi^2 EA}{(l/r_Y)^2} = 0.6 P_{EY}$$

If we now introduce a fictitious effective length αl then we can describe the torsional–flexural buckling load in terms of the lower Euler critical load P_{EY} as

$$P_{cr} = \frac{\pi^2 EA}{(\alpha l/r_Y)^2} \tag{2.32}$$

where in this case $\alpha = 1.29$.

Values for α corresponding to various geometries have been determined[30] and examples are shown in Figures 2.28 and 2.29. Values for lipped and plain channel sections with equal flanges are tabulated in BS 449 Addendum 1,

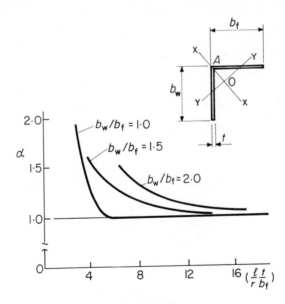

Figure 2.28

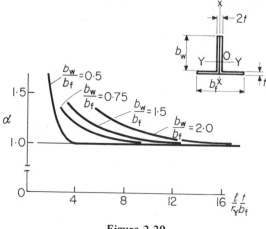

Figure 2.29

and it is worth while to note here that for these sections, provided the overall width of the web is greater than, or equal to, 1.3 times that of the flange, the value of α may be taken as unity.

So far in this section we have considered only the critical loads; these of course are the loads at which, if the strut were perfectly straight and the load perfectly aligned with the centroidal axis, the strut would first begin to buckle. But as we discussed earlier in this chapter, no such perfect strut can exist in practice and we must introduce some form of imperfection, for example an initial deflection. If, for torsional–flexural buckling, the variation of the imperfection magnitude e is taken to be similar to that which is commonly used for the Euler flexural buckling of hot-rolled sections, that is

$$\eta \equiv \frac{ec}{r^2} = 0.3 \times 10^{-4} \left(\frac{\alpha l}{r}\right)^2 \qquad \text{(see equation 2.5)}$$

the maximum load P_{mE}, given by the Perry formula, equation 2.3, can be calculated using

$$\frac{P_{mE}}{P_y} = \frac{1}{2}\left[1 + (1+\eta)\frac{P_E}{P_y}\right] - \sqrt{\left\{\left[1 + (1+\eta)\frac{P_E}{P_y}\right]^2 - \frac{P_E}{P_y}\right\}} \qquad (2.33)$$

where P_y is the crushing load and

$$\frac{P_E}{P_y} = \frac{\pi^2 E}{\sigma_y} \frac{1}{(\alpha l/r)^2}$$

This is shown in Figure 2.30 and we see one limit is $P_{mE} \rightarrow P_y$ as $\alpha l/r \rightarrow 0$. In using this analysis we are assuming that imperfections have the same effect on

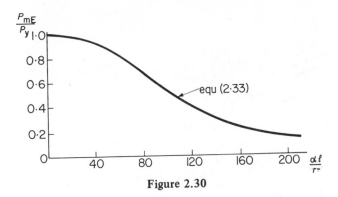

Figure 2.30

torsional—flexural buckling as we know, from theory and experiment, that they have for purely flexural buckling. But at the geometries for which the torsional—flexural and the purely flexural critical loads coincide, it is possible that some interaction will occur between the two modes which could result in a reduction in the maximum load the strut can carry. There is little theoretical information on this subject and some research is needed.

2.4 INTERACTION BETWEEN LOCAL AND OVERALL BUCKLING

In the derivation of equation 2.33 it has been assumed that as the slenderness ratio $\alpha l/r$ is reduced the maximum load that can be carried by the column will approach the crushing load. But from Section 2.2 we know that due to local buckling the maximum load that can be carried by a strut that has not experienced overall buckling will be less than the crushing load. The theoretical problem of determining the behaviour of a strut that is experiencing local and overall deformations is very difficult. It involves knowing the pattern of local imperfections and being able to determine the local stiffness of the section and thence the overall stiffness of the strut. From this latter information it is then possible to determine the magnitude of the overall deflections, relative to the initial overall deflection, and the load corresponding to these deflections. A numerical solution has been presented for closed rectangular box sections.[31]

A much simpler, although not so securely founded, approach is to assume that since the short strut will collapse at the maximum load P_{mL} we can use this load in equation 2.33 instead of the crushing load P_y. Hence we have

$$\frac{P_{mE}}{P_y} = \frac{1}{2}\left[\frac{P_{mL}}{P_y} + (1+\eta)\frac{P_E}{P_y}\right] - \sqrt{\left\{\frac{1}{4}\left[\frac{P_{mL}}{P_y} + (1+\eta)\frac{P_E}{P_y}\right]^2 - \frac{P_{mL}}{P_y}\frac{P_E}{P_y}\right\}} \qquad (2.34)$$

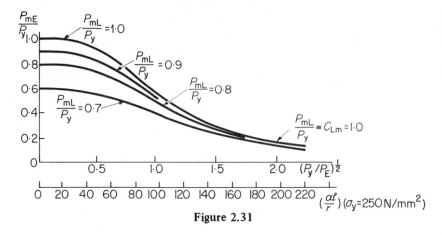

Figure 2.31

Equation 2.34 is shown in Figure 2.31 for a number of values of section local effectiveness C_{Lm}, that is P_{mL}/P_y. This analytical approach is based empirically on a number of tests,[32] example results of which are shown in Figure 2.32. Values of the maximum average allowable stress p_a (that is the maximum stress

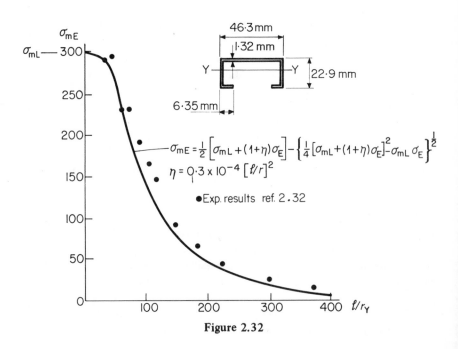

$$\sigma_{mE} = \frac{1}{2}\left[\sigma_{mL} + (1+\eta)\sigma_E\right] - \left\{\frac{1}{4}\left[\sigma_{mL} + (1+\eta)\sigma_E\right]^2 - \sigma_{mL}\,\sigma_E\right\}^{\frac{1}{2}}$$

$$\eta = 0.3 \times 10^{-4}\left[\ell/r\right]^2$$

●Exp. results ref. 2.32

Figure 2.32

divided by a safety factor of 1.7) are tabulated in BS 449 Addendum 1 for various values of C_{Lm} and slenderness ratios $\alpha l/r$; the values of p_a have been calculated using equation 2.34. (Notice that the local effectiveness C_{Lm} in that table is related to a steel having $\sigma_y = 250$ N/mm^2.)

Example 2.2
Now let us calculate the maximum load for a strut having the cross-section in Figure 2.24 and an effective length of 2.5m. From page 52 we have the maximum local stress as $\sigma_{mL} = 320$ N/mm^2, thus $P_{mL}/P_y = 0.92$. Ignoring the effect of corners we can represent the cross-section in Figure 2.24 approximately and the position of the centroid is calculated from the first moment of area as

$$2 \times (17 \times 3 \times 48.5) + 2 \times (50 \times 3 \times 25) + 94 \times 3 \times 1.5$$
$$= y \left[2 \times 17 \times 3 + 2 \times 50 \times 3 + 94 \times 3 \right]$$

Notice that since the thickness is constant around the cross-section it can be eliminated from the above equation. Hence, $\bar{y} = 18.8$ mm and the approximate (ignoring terms involving the square of the thickness) second moment of area about YY is

$$I_Y = 94 \times 3 \times (18.8 - 1.5)^2 + 2 \times \frac{1}{12} \times 3 \times 50^3 + 50 \times 3 \times (25 - 18.8)^2 \times 2$$
$$+ 2 \times 17 \times 3 \times (50 - 1.5 - 18.8)^2$$
$$= 2.43 \times 10^5 \text{ mm}^4$$

The cross-sectional area A is
$$A = 684 \text{ mm}^2$$

The web/flange ratio is $\simeq 2$, which obviously is in excess of 1.3, hence the section is not prone to torsional–flexural instability, thus $\alpha = 1$.

$$r_Y = \sqrt{\left(\frac{I_Y}{A}\right)} = \left(\frac{2.43 \times 10^5}{684}\right)^{\frac{1}{2}} = 18.9 \text{ mm}; \frac{\alpha l}{r_Y} = 132.6; \sigma_E = \frac{\pi^2 E}{(\alpha l/r_Y)^2} = 112.3 \text{ N/mm}^2$$

Hence from equation 2.34, or Figure 2.31, we have $P_{mE}/P_y = 0.25$. Thus the permissible load P_a is

$$P_a = \frac{1}{1.7} \times 0.25 \times (684 \times 350) = 35.9 \text{ kN}$$

An alternative design procedure is to use the tabulated information provided in British Standards. Thus from page 53 we have the approximate local effectiveness $C_{Lm} = 0.99$ but in order to use Table 6 in BS 449 Addendum 1 we require the corresponding value for $\sigma_y = 350$ N/mm^2. The corrected value is

$$C_{Lm} = \frac{350}{250} \times 0.99 = 1.39$$

From BS 2994 we have area = 650 mm^2, $r_Y = 18.7$ mm, and from BS 449 Addendum 1, Table 6, $p_a = 53.1$ N/mm^2, hence

$$P_a = 53.1 \times 650 = 34.5 \text{ kN}$$

2.5 REFERENCES

1 Euler, L., *Isis*, Vol. 20, No. 58, p. 1, 1933.
2 Walker, A. C., *Buckling of Struts*, Chatto & Windus, 1975.
3 Croll, J. G. A. and Walker, A. C., *Elements of Structural Stability*, MacMillan 1972.
4 Perry, I., *The Engineer*, Vol. 62, p. 646, 1886.
5 Robertson, A., *Selected Engineering Papers, No. 28*, Institution Civil Engineers, 1925.
6 Godfrey, G. B., *The Structural Engineer*, Vol. 40, p. 97, 1962.
7 Dutheil, I., Preliminary Publication, 4th Congress IABSE, p. 275, 1952.
8 Bulson, P. S., *The Stability of Flat Plates*, Chatto & Windus, 1970.
9 Specification for the Design of Cold-formed Steel Structural Members, American Iron and Steel Institute, 1968, New York.
10 Coan, J. M., *Trans. ASME, J. App. Mech.*, Vol. 18, p. 143, 1951.
11 Walker, A. C., *Aero. Quart.* Vol. XX, p. 203, 1969.
12 Dawson, R. G. and Walker, A. C., *Proc. ASCE, J. Struct. Div.*, Vol. 98, p. 75, 1972.
13 Graves-Smith, T. R., Proceedings of the International Conference on Structures Solid Mechanics and Engineering Design, J. Wiley, 1969.
14 Stowell, E. Z., NACA Report 1029, 1951.
15 Kenedi, R. M. and Harvey, J. M., *Trans. Inst. Eng. Shipbldrs. Scot.*, Vol. 94, p. 109, 1950–51.
16 Rhodes, J. and Harvey, J. M., *Int. J. Mech. Sci.*, Vol. 13, p. 787, 1971.
17 Rhodes, J. and Harvey, J. M., *J. Mech. Eng. Sci.*, Vol. 13. p. 867, 1971.
18 Walker, A. C. *Proc. ASCE, J. Struct. Div.*, Vol. 92, p. 39, 1966.
19 Harvey, J. M., Ph.D. Thesis University of Glasgow, 1952.
20 Harvey, J. M., *Engineering*, Vol. 75, p. 291, 1953.
21 Chilver, A. H., *Aero. Quart.*, Vol. 4, p. 251, 1953.
22 Chilver, A. H., *The Engineer*, p. 180, Aug. 7, 1953.
23 Winter, G., *Trans. ASCE*, Vol. 112, p. 527, 1947.
24 Winter, G., Preliminary Publications 3rd Congress IABSE, p. 237, 1948.
25 Bulson, P. S., *Int. J. Mech. Sci.*, Vol. 11, p. 613, 1969.
26 Reiss, M. and Chilver, A. H., 8th Congress, IABSE, 1968.
27 Timoshenko, S. P., *Theory of Elastic Stability*, McGraw-Hill, 1961.

28 Timoshenko, S. P., *J. Franklin Inst.,* Vol. 239, Nos. 2, 3 and 4, 1945.
29 Chilver, A. H., *Proc. ICE,* Vol. 20, p. 233, 1961.
30 Kenedi, R. M. and Chilver, A. H., Institute of Sheet Metal Engineering
    Symposium on the Application of Sheet and Strip Metals in Building, 1959.
31 Graves-Smith, T. R., 8th Congress, IABSE, 1968.
32 Chilver, A. H., Ph.D. Thesis, University of Bristol, 1950.

3

Beam Bending

3.1 INTRODUCTION

Engineer's bending theory predicts that the stresses and deflections of a loaded beam are proportional to the applied load. It has long been known, however, that cold-formed flexural members often do not behave as simple beam theory indicates. The flexural rigidity of such beams can be substantially less at high loads than it is at low loads, and the predicted stresses calculated using simple beam theory are sometimes much lower than those actually present in these beams. In the design of cold-formed beams, these effects must be taken into account in order to ensure safe and efficient design. For the complete design of a cold-formed beam it is necessary to include the stresses arising from the torsional loading and this is treated in Chapter 4.

Many of the problems encountered in cold-formed beam design arise from the thinness of material used in their manufacture. The discussion in the last chapter showed that thin sheeting has a tendency to buckle out of plane in the presence of compressive stresses. This out-of-plane buckling causes substantial variation in the behaviour under further load.

Figure 3.1 shows a cold-formed channel section under moment. The compression element has developed buckles along its length which reduce its stiffness and carrying capacity. This type of buckling is classified as *local buckling of the compression flange*. In order to use cold-formed beams efficiently, it is sometimes desirable to design them for use in the post-buckled state and to do this it is necessary to apply the theory of plate buckling developed in Chapter 2.

Another form of local instability which is encountered is *web buckling*. This can occur near points of concentrated load or at support points. A sketch of a typical web buckling condition is shown in Figure 3.2; the web has buckled under the very high local compressive stresses imparted to it by the support. Unlike compression flange buckling, the beam is unable to function usefully after web buckling, and so this type of local instability must be avoided.

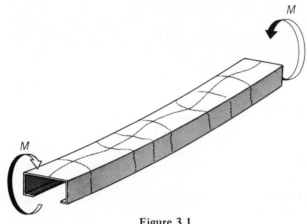

Figure 3.1

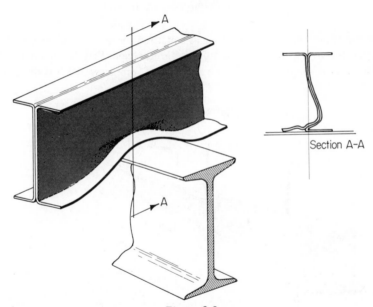

Section A-A

Figure 3.2

As with hot-rolled sections, *shear buckling* of deep thin webs must also be avoided. The shear stresses in a beam carrying variable bending movement can cause the webs to buckle as shown in Figure 3.3.

Apart from local buckling, care must be taken in the design of long beams which are bent about their stronger principal axis, for there is often a tendency for the beam to bend also about its weaker axis when some critical stress is

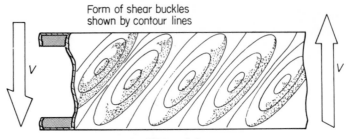

Figure 3.3

reached, as shown in Figure 3.4. This type of overall buckling is termed *lateral buckling.*

In this chapter, these effects (with the exception of web buckling) will be discussed in greater detail.

3.2 ENGINEER'S THEORY OF BENDING

One of the advantages of cold forming as a method for manufacturing structural sections is the very wide variety of cross-sectional shapes that can be obtained. In general, these shapes may not have any axes of symmetry and it is therefore

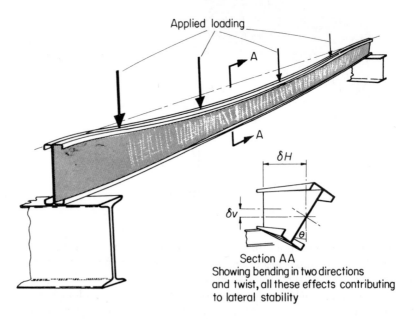

Section AA
Showing bending in two directions
and twist, all these effects contributing
to lateral stability

Figure 3.4

useful at this stage to review the theory of *bending of unsymmetrical sections*
that is commonly used in engineering analysis. This theory is based on the
assumption that initially plane sections remain plane after the beam has been
loaded. It predicts, with an accuracy sufficient for engineering purposes, the
deflections and stress distributions for beams whose lengths are much greater
than the cross-sectional dimensions. For short beams, shear stresses, induced
by loading or varying bending moments, become more important and the simple
theory developed here is less applicable. Also, the analysis, by assuming the cross-
section to remain unaltered in shape during loading, takes no account of buckling.
This effect is discussed later in this chapter.

Consider a cold-formed beam with a general cross-section shape, for example
Figure 3.5, we can set up an arbitrary set of axes OX, OY, OZ, where the last
is directed along the centre line of the beam. It is common for O to coincide
with the centroid of the section since it will then lie on the neutral axis which
always passes through the centroid. Suppose the section is subjected to a
bending moment about some axis AOA, we can resolve this into two
components, one moment M_Y, about the OY axis and another M_X, about
the OX axis. The relationship between the moments is given by

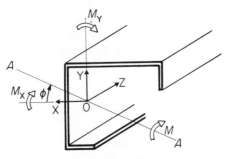

Figure 3.5

$$M_X = M \cos \phi; M_Y = M \sin \phi \qquad (3.1)$$

The application of the moment results in direct stress in the longitudinal direction
being induced across the section. It may be shown[1] that at a point on the cross-
section distance x and y from O the stress is

$$\sigma_z = -\frac{M_X{}^*}{I_X} y + \frac{M_Y{}^*}{I_Y} x \qquad (3.2)$$

where, as before, tensile stresses are positive. The components $M_X{}^*$ and $M_Y{}^*$ are sometimes called the *effective moments* and are defined by

$$M_X{}^* \equiv \frac{M_X + M_Y\,I_{XY}/I_Y}{1 - I_{XY}{}^2/I_X\,I_Y} \; ; \; M_Y^* \equiv \frac{M_Y - M_X\,I_{XY}/I_X}{1 - I_{XY}{}^2/I_X\,I_Y} \tag{3.3}$$

The geometric constants I_X and I_Y are the second moments of area about OX and OY respectively, and I_{XY} is the product moment of area defined by

$$I_{XY} = \int_A x\,y\,\mathrm{d}A \tag{3.4}$$

It is important to realize that if OX or OY coincides with an axis of symmetry this product moment of area is zero, and $M_X{}^* = M_X \,; M_Y{}^* = M_Y$. For every cross-sectional shape there is a neutral axis along which the stress is zero. Setting $\sigma_z = 0$ in equation 3.2 we have the angle θ which this axis makes with the OX axis,

$$\theta = \tan^{-1}\frac{M_Y{}^*I_X}{M_X{}^*I_Y} \tag{3.5}$$

where θ is measured positively in a clockwise direction from OX (see Figure 3.6). Since the basic assumption for this theory is that plane sections remain plane, the beam will deflect by sections rotating about the neutral axis. Thus,

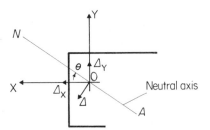

Figure 3.6

if Δ_X and Δ_Y are the deflections at some distance z along the beam in the X and Y directions, respectively, they are related to the deflection Δ normal to the neutral axis (see Figure 3.6) by,

$$\Delta_X = \Delta \sin\theta\,; \Delta_Y = -\Delta\cos\theta$$

Further, from the expression for the stress distribution (equation 3.2) and an approximation that the curvature normal to the neutral axis is given by $d^2 \Delta/dz^2$, we have, in a similar way to the more usual treatment of Engineer's bending theory,

$$\frac{d^2\Delta_X}{dz^2} = \frac{M_Y{}^*}{EI_Y}; \quad \frac{d^2\Delta_Y}{dz^2} = -\frac{M_X{}^*}{EI_X} \tag{3.6}$$

Example 3.1
The cold-formed angle section shown in Figure 3.7 is to be used for a simply supported beam 2 m long. It carries a vertical concentrated load of 0.5 kN at

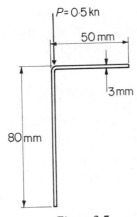

Figure 3.7

mid-length in line with the web. The first task is to calculate the position of the centroid.

$$A = 3 \times 77 + 50 \times 3 = 381 \text{ mm}^2$$

$$A \times \bar{x} = (77 \times 3 \times 1.5) + (50 \times 3 \times 25); \bar{x} = 10.8 \text{ mm}$$

$$A \times \bar{y} = (77 \times 3 \times 41.5) + (50 \times 3 \times 1.5); \bar{y} = 25.8 \text{ mm}$$

$$I_X = (\frac{1}{12} \times 50 \times 3^3) + 50 \times 3 \times (24.3)^2 + \frac{1}{12} \times 3 \times 77^3 + 77 \times 3 \times (15.7)^2$$

$$= 112 + 88\ 573 + 114\ 133 + 56\ 939 = 25.98 \times 10^4 \text{ mm}^4$$

Notice that if we had neglected the first contribution to I_X we would have made little difference to the accuracy of the result. It is generally permissible, and certainly worth while from the viewpoint of reducing the amount of calculation in complex cross-sections, to ignore those terms that involve the square, and higher orders, of the thickness. Also, later in the calculation we shall ignore the variation of direct stress across the thickness of the material and carry out the analysis using the centre line dimensions.

$$I_Y = 77 \times 3 \times (9.3)^2 + \frac{1}{12} \times 3 \times 50^3 + 50 \times 3 \times (14.2)^2 = 8.15 \times 10^4 \text{ mm}^4$$

$$I_{XY} = 50 \times 3 \times 24.3 \times (-14.2) + 77 \times 3 \times 9.3 \times (-15.7) = -8.55 \times 10^4 \text{ mm}^4$$

In this example, $M_Y = 0$ and at the centre of the beam

$$M_X = \frac{Pl}{4} = \frac{0.5 \times 2}{4} = 0.25 \text{ kNm}$$

Hence,

$$M_X{}^* = \frac{M_X}{1 - (8.55)^2/(25.98 \times 8.15)} = \frac{M_X}{1 - 0.345} = 0.382 \text{ kN m}$$

and

$$M_Y{}^* = -0.126 \text{ kN m}$$

From equation 3.2 the stress at some point x, y on the centre of the cross-section at midlength of the beam is

$$\sigma_z = \frac{-0.382 \times 10^6}{25.98 \times 10^4} \, y - \frac{0.126 \times 10^6}{8.15 \times 10^4} x = (-1.47 \, y - 1.55 \, x) \text{ N/mm}^2$$

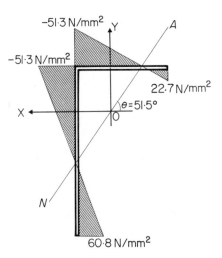

Figure 3.8

At the corner for example, $x = 9.3$ mm, $y = 24.3$ mm and therefore $\sigma_z = -51.3$ N/mm^2. The stress distribution is shown in Figure 3.8 and the neutral axis inclination is determined by

$$\theta = \tan^{-1}\left[\frac{-0.126 \times 25.98}{0.382 \times 8.15}\right] = -51.5^0$$

From equation 3.6

$$\frac{d^2\Delta_X}{dz^2} = \frac{M_Y{}^*}{EI_Y}$$

but

$$M_Y{}^* = M_X\left[\frac{I_{XY}/I_X}{1 - I_{XY}{}^2/I_X I_Y}\right]$$

and

$$M_X = \frac{P}{2}z$$

where P is the concentrated load and z is the distance along the beam from one of the supports. Hence,

$$\frac{d^2\Delta_X}{dz^2} = \frac{1}{EI_Y}\left[\frac{I_{XY}/I_X}{1 - I_{XY}^2/I_X I_Y}\right]\frac{Pz}{2} \tag{3.7}$$

Equation 3.7 has a form similar to the usual equation governing the deflection of a beam,

$$\frac{d^2\Delta}{dz^2} = -\frac{M}{EI}$$

where M is the moment at some section distance z from one end of the beam. In this particular example we have a beam subject to a central point load P and, if we have simple supports at the ends, the deflection Δ at the beam mid-length will be

$$\Delta = \frac{Pl^3}{48EI}$$

Similarly if we integrate equation 3.7 with the appropriate simple support conditions at the ends, the deflection Δ_X at mid-span in the X direction will be given by

$$\Delta_X = \frac{P_X{}^*l^3}{48EI_Y},$$

where

$$P_X{}^* \equiv P \ \frac{I_{XY}/I_X}{1 - I_{XY}{}^2/I_X I_Y}$$

Similarly the deflection in the Y direction is

$$\Delta_Y = - \frac{P_Y{}^* l^3}{48 E I_X},$$

where

$$P_Y{}^* \equiv P \ \frac{1}{1 - I_{XY}{}^2/I_X I_Y}$$

Comparison with equation 3.3 suggests that $P_X{}^*$ and $P_Y{}^*$ could be called *effective loads* and that loading in any plane could be treated by separation into components along the chosen axes OX and OY.
In this example

$$P_X{}^* = \frac{0.5 \times -8.55/25.98}{1 - 0.345} = -0.25 \text{ kN}$$

$$P_Y{}^* = \frac{0.5}{1 - 0.345} = 0.76 \text{ kN}$$

$$\Delta_X = \frac{(-0.25) \times 8 \times 10^9}{48 \times 200 \times 8.15 \times 10^4} = -2.5 \text{ mm}$$

$$\Delta_Y = \frac{-0.76 \times 8 \times 10^9}{48 \times 200 \times 25.98 \times 10^4} = -2.4 \text{ mm}$$

Hence, although the load is applied vertically, we have a horizontal deflection greater than the deflection in the direction of the load. This is due to the asymmetry of the section. If, as is common in some commercial cold-formed sections, there is a plane of symmetry, and if the beam is loaded only in that plane, the deflection will only be in the direction of the load. Also, beams are usually attached to the loading member, possibly a floor or roof sheet, and the attachments will tend to restrict the deflection normal to the line of loading. However, it must be remembered that this may cause the beam to twist and that the attachments must be strong enough to resist the sideways movement.

3.3 DESIGN OF THIN WALLED BEAMS

The previous section outlined the type of analysis that must be used if we are to design an asymmetric beam. The theory was based on the assumption that plane sections remain plane and that the cross-section is undistorted. This implies that it can take no account of the local deformations that inevitably occur in thin walled beams. In this section we introduce a simple design approach that accounts for these local effects but because of the complexity of analysing a beam with any general cross-section and loading we restrict attention to beams having one axis of symmetry and loaded in the plane parallel to that axis.

3.3.1 EFFECTS OF BUCKLING ON SECTION PROPERTIES

The cross-section of a beam subject to moment loading is composed of tension elements, compression elements and web elements. As an example, the lipped channel section shown in Figure 3.9 (a) has a compression element①, two tension elements③ and two elements②which will withstand the shear resulting from varying bending movement. On elements② the direct stress varies from compression at the junction with①to tension at the junction with③ as shown in Figure 3.9 (b). At a point on this element, the stress passes through zero. This point is coincident with the neutral axis of bending and is located such that the net longitudinal force on the section is zero. The distance $\bar{y}$ to the neutral axis may be determined from the equation

$$\bar{y} = \frac{\Sigma Ay}{\Sigma A},$$

where ΣAy is the first moment of area about OO and ΣA is the cross-sectional area.

Now, if moments are applied to the beam of such magnitude that the compressive stress on element① is greater than the critical stress, then this element buckles and becomes less effective. The reduction in effectiveness of the compression element can be evaluated using the C_L $(\equiv P_{mL}/P_y)$ factors for any particular b/t ratio. Also, using the effective width concept, the stress distribution on the beam can be idealized as shown in Figure 3.10. Since the whole of the compression element is not now effective, then ΣA is effectively

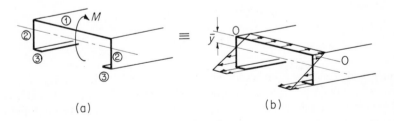

(a)　　　　　　　　　　　　(b)

Figure 3.9

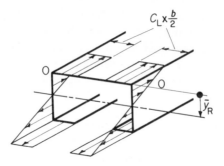

Figure 3.10

reduced. Because of this, the value of $\bar{y}$ and the position of the neutral axis change. Therefore, in the post-buckling range, the compression elements lose part of their effectiveness, causing changes in the position of the neutral axis and decreasing the I and Z values of the section, where Z is the *section modulus,* defined as the second moment of area divided by the distance from the neutral axis to the outermost point on the cross-section.

3.3.2 COMPUTATION OF SECTION PROPERTIES AND REDUCED SECTION PROPERTIES

In the evaluation of section properties for use in design the effects of local buckling must therefore be included. For compression elements parallel to the axis about which bending occurs, such as elements ① in Figure 3.11, the

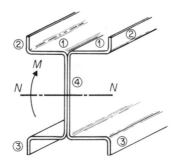

Figure 3.11

reduction in efficiency is taken into account using the C_L values derived in Chapter 2, curve A in Figure 2.16, and also given in Table 5 of BS 449 Addendum 1. For compression elements perpendicular to this axis—such as the lip elements ②—no analytical information is available. This type of element is dealt with by a more empirical approach. Observation of the section in Figure 3.11 shows that the lips② are in compression, which increases from the supported edge to the unsupported edge. If this type of element is very wide, the load condition is extremely severe from a buckling point of view, and the effectiveness of the lips will be very small. Therefore, for safe design, the wisest course is to ignore completely the effects of outwardly turned lips, except for their stiffening effect on the compression flanges if their width is sufficiently large. For steel with yield stress of 250 N/mm², the limiting width of this type of lip, beyond which its contribution to the effective section properties should be ignored, is 10*t*. Nevertheless, it remains a requirement that the lip width must at least equal one-fifth of the adjacent flange width to provide adequate edge support.

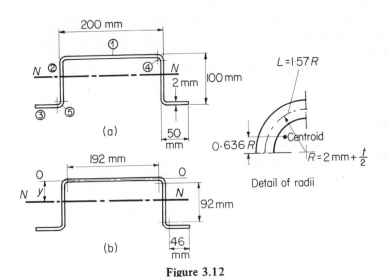

Figure 3.12

If, even with this requirement, the lip width is less than 10*t*, it can be considered that buckling will not be involved when the lip is loaded. Therefore the whole lip area may be included in the section property calculations.

For steels with yield stresses other than 250 N/mm², the limiting width can be calculated from $10t \sqrt{(250/\sigma_y)}$.

If the elements② are turned inwards towards the neutral axis, the compressive stress due to bending decreases towards the unsupported edges, so that buckling becomes less likely. In this case, the whole area of the lips can safely be used in determination of section properties. When considering lips on the tension side of the section, such as elements③ the problem of buckling does not arise so that the whole area of these lips should be included in calculations. In computing the properties of a section with thin walls, calculation can be reduced if these properties are computed in terms of element lengths rather than area, with the thickness t kept as a multiplying factor; that is, ignoring contributions involving the square of the thickness. If it is required to compute both full section properties and reduced section properties (as is the case when deflection determination is required), further labour saving can be achieved by taking the reference axis through the compression elements. An example is shown here to illustrate these points.

Example 3.2
Figure 3.12 shows a top hat section which is to be bent about axis N–N in such a manner that the top element ① is in compression. It is required to evaluate the section properties and reduced section properties for steel of yield stress 250 N/mm^2.

The reference axis O–O is taken through the centre line of the compression element ①. The properties are then tabulated as in Table 3.1.

Table 3.1

Element	b (mm)	y (mm)	by (mm$^2 \times 10^2$)	by^2 (mm$^3 \times 10^3$)	I_{cg}/t (mm$^3 \times 10^3$)	I_{00}/t (mm$^3 \times 10^3$)
①	192	0	0	0	≈ 0	≈ 0
2 x②	184	50	92	460	130	590
2 x③	92	98	90.2	884	≈ 0	884
2 x④	4.7	2.1	0.1	≈ 0	≈ 0	≈ 0
2 x⑤	4.7	97.9	4.6	45	≈ 0	45
$\Sigma b = 477.4$			$\Sigma by = 186.9$			$\Sigma I_{00}/t = 1519$

$$\bar{y} = \frac{\Sigma by}{\Sigma b} = \frac{186.9 \times 10^2}{477.4} = 39.3 \text{ mm}$$

$$I_{NA} = t \times (I_{00}/t - \Sigma b\bar{y}^2) = 7.2\,(1519 \times 10^3 - 477.4 \times 39.3^2) = 155.4 \times 10^4 \text{ mm}^4$$

Distance from neutral axis to extreme fibres:

(1) in compression $= (y + 1) \text{ mm} = 39.3 + 1 = 40.3 \text{ mm}$
(2) in tension $= 100 - 40.3 = 59.7 \text{ mm}$

$$Z_{(min)} = \frac{155.4 \times 10^4}{59.7} = 26 \times 10^3 \text{ mm}^3$$

For reduced section, if the yield stress is 250 N/mm², then from Figure 2.16 the C_L value of element ① is 0.406. Thus Σb becomes $477.4 - 192 + (0.406 \times 192) = 363.3$ mm. Note that the other quantities in Table 3.1 remain unaltered because of the choice of reference axis. Therefore the new value of $\bar{y}$ is

$$\bar{y}_R = \frac{\Sigma by}{\Sigma b} = \frac{186.9 \times 10^2}{363.3} = 51.4 \text{ mm}$$

$$I_R = 2 \times (1519 \times 10^3 - 363.3 \times 51.4^2) = 112 \times 10^4 \text{ mm}^4$$

$$Z_{R(min)} = \frac{112 \times 10^4}{51.4 + 1} = 21.4 \times 10^3 \text{ mm}^3$$

Now all relevant bending properties are known for the section, and analysis of design loads and deflections can proceed. It is evident in this example that the computation is eased without undue loss of accuracy if the section is considered to have sharp corners; indeed the effects of the small radii are minute.

3.3.3 COMPUTATION OF PERMISSIBLE DESIGN LOADS

Under working conditions, it is desirable that a loaded beam operates with a known factor of safety on the collapse load. It has been found experimentally that collapse of a locally buckled beam occurs when the maximum stress on the compression element reaches yield stress. Therefore, a reasonably accurate estimate of the collapse load will be obtained if the maximum moment is computed from the simple bending formula, altered to take reduction in effectiveness into account,

$$M_c = \sigma_y Z_R$$

where M_c is the collapse moment, Z_R is the relevant reduced modulus for the section and σ_y is the yield stress of the material. It should be understood that the effectiveness of the compression element (and hence of the section) reduces with increased stress, so that at stresses below yield, the efficiency of the section is greater than at yield. Therefore, if a true safety factor based on stress were required for design, one would have to evaluate C_L factors based on permissible design stress. This could be done if necessary but, as previously stated, it is felt that a safety factor based on collapse load is preferable since the degree of safety ensured is then more easily visualized. On these grounds, BS 449 Addendum 1 indicates that the fully reduced modulus should be used in computation of permissible design loads with a factor of safety. The maximum allowable moment M_a on the section is therefore

$$M_a = 0.65 \, (\sigma_y Z_R)$$

which gives a safety factor of 1/0.65 on the collapse moment. The safety factor on stress would consequently be slightly greater than this owing to the factors already discussed, but this is a comforting thought rather than a cause for concern.

Figure 3.13 shows a comparison[2] between the permissible design moments for lipped channel sections based on the above equation and the results of tests carried out to collapse on these sections. The experimental collapse loads have

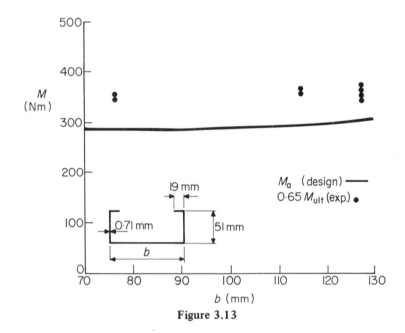

Figure 3.13

been multiplied by 0.65 to facilitate comparison and it can be seen that the theoretical results are slightly conservative, owing to the fact that the C_L values were computed for simply supported elements whereas the compression elements in this case have some rotational restraint from the webs. Again, these results confirm the safety of the design criteria.

Applying the computational procedure to the section discussed in Section 3.3.2 gives, for the permissible moment based on a yield stress of 250 N/mm^2,

$$M_a = 0.65 \times 250 \times 21.4 \times 10^3 = 3.48 \times 10^6 \text{ N mm}$$

that is,

$$M_a = 3.48 \text{ kN m}$$

3.3.4 CALCULATION OF DEFLECTIONS

Since a beam will undergo part of its loading in an unbuckled state, and part in
a buckled state, the load/deflection relationship does not follow a straight line.
A general theoretical load displacement curve for a cold-formed beam is shown
in Figure 3.14. The gradient of the curve decreases after buckling and decreases
still further near the collapse load when some plasticity may be present. The
condition near the collapse load is of little interest for design since this is to be
avoided by the reduction to allowable moment.

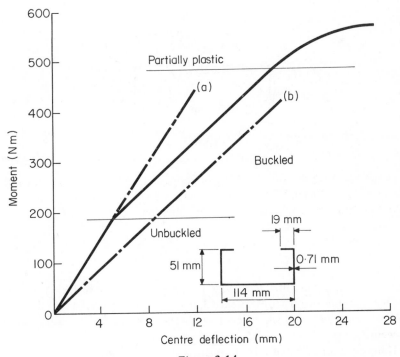

Figure 3.14

In evaluation of the deflection for such a beam, the question arises: which
section properties should be used? Using the full second moment of area predicts
deflections as shown by line (a) in Figure 3.14, which seriously underestimates
the real deflections. Using the reduced second moment of area gives line (b) which
overestimates the deflections considerably and will cause over-conservatism if
design is governed by deflection. Therefore, if the displacements are required
with reasonable accuracy, some attempt must be made to give consideration to
both pre-buckling and post-buckling effects on deflection behaviour.

In view of this the moment required to cause buckling must be known. It can be calculated using Engineer's beam theory together with the critical load coefficient k (see equation 2.9) for the compression element. Now, the use of $k = 4$ (the value for a simply supported plate) for stiffened plates which form part of flexural members is sometimes unduly conservative if the flange has substantial webs at both sides. These webs, which are less liable to buckling than the compression element due to their partly compressive/partly tensile stress distribution, impose a degree of clamping on the compression element, as shown in Figure 3.15 for a lipped channel-section beam, and thus raise the element's resistance to buckling.

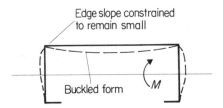

Figure 3.15

A rigorous analysis of the buckling of lipped-channel beams[3,4] has shown that the buckling coefficients for compression elements in such a beam can be expressed with reasonable accuracy by the equation

$$k = 5.23 + 0.16 \frac{b_w}{b_f} \tag{3.8}$$

This represents a substantial increase in buckling strength for such elements and, judging from the analysis in Reference 3, may be considered valid for all symmetrical beams in which the compressive elements have both edges supported by webs.

Using equations 3.8 and 2.16, the critical stress σ_{cr} of the compression element can be evaluated. The critical moment is found from

$$M_{cr} = \sigma_{cr} Z$$

where Z in this case is the full section modulus, since this reduces only after buckling.

In BS 449 Addendum 1, the critical stress is found from the formula

$$\sigma_{cr} = \frac{12 \times 10^4 k}{(b_w/t)^2} \ \text{N/mm}^2$$

The number 120 000 was arrived at by taking $E = 207$ kN/mm^2 and dividing by 1.5 in equation 2.9. This gives a safety factor of 1.5 on the buckling load of the beams under consideration to allow for any imperfections in shape or initial deformations which may be present.

The deflections up to the critical load may be evaluated using simple beam theory and the full section properties of the beam, since the beam is fully effective before buckling. For computation of the deflection Δ_a at the allowable load, the fully reduced section is used. Therefore Δ_{cr} and Δ_a can now be plotted as shown in Figure 3.16. Obviously the load/deflection curve for moments less than M_{cr} takes the form of a straight line joining Δ_{cr} and the origin. For the evaluation of deflections at loads greater than M_{cr} the assumption is made that the load/deflection curve in this region too consists of a straight line joining

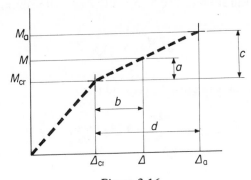

Figure 3.16

Δ_{cr} and Δ_a. This assumption is only approximate since, in reality, the line joining these two points is curved, but the errors introduced in the approximation are acceptable and conservative. Therefore, for moments between M_a and M_{cr} the displacement can be found in the following way: from similar triangles in Figure 3.16, we have

$$\frac{a}{b} = \frac{c}{d}, \text{ or } \frac{a}{c} = \frac{b}{d}$$

but, since $a \equiv M - M_{cr}$, $b \equiv \Delta - \Delta_{cr}$, $c \equiv M_a - M_{cr}$, $d \equiv \Delta_a - \Delta_{cr}$ we have

$$\frac{M - M_{cr}}{M_a - M_{cr}} = \frac{\Delta - \Delta_{cr}}{\Delta_a - \Delta_{cr}}$$

Since only two unknowns M and Δ are present then for a specified moment M the corresponding deflection Δ can be obtained. Alternatively, if Δ is specified corresponding moment M can be obtained.

Example 3.2 (continued)
As an illustration of the method, consider that a beam with the section discussed in Section 3.3.2 is loaded by a uniformly distributed load w/unit length over a span of 6 m, with simple supports at both ends. The critical moment on the beam is found from

$$M_{cr} = \frac{12 \times 10^4 \times (5.23 + 0.16\, b_w/b_f) \times Z_c}{(b_w/t)^2} = 1.8 \text{ kN m}$$

The maximum allowable moment on the beam is found from

$$M_a = 0.65 \times 250 \times Z_R = 3.48 \text{ kN m}$$

for a beam with yield stress of 250 N/mm^2. The value of w to cause M_{cr} is found from

$$w_{cr} = \frac{8M_{cr}}{l^2} = 0.4 \text{ kN/m}$$

Similarly, the value of w to cause M_a is found from

$$w_{cr} = \frac{8M_a}{l^2} = 0.744 \text{ kN/m}$$

The deflection at w_{cr} is

$$\Delta_{cr} = \frac{5\, w_{cr} l^4}{384\, EI}$$

where I is the full second moment of area. Taking $E = 200$ kN/mm^2, we have $\Delta_{cr} = 21$ mm. The deflection at w_a is

$$\Delta_a = \frac{5\, w_a\, l^4}{384E\, I_R} = 56.5 \text{ mm}$$

If the deflection is required under a working load of 0.6 kN/m, for which $M = 2.7$ kN m, rearrangement of equation 3.8 gives

$$\Delta = \Delta_{cr} + \frac{M - M_{cr}}{M_a - M_{cr}} (\Delta_a - \Delta_{cr})$$

But,

$$M = \frac{wl^2}{8} = 2.7 \text{ kN m, hence } \Delta = 21 + \frac{2.7 - 1.8}{3.48 - 1.8} (56.5 - 21) = 40 \text{ mm}$$

Conversely, if deflections govern design, the maximum permissible load for a deflection limitation, of $l/240$ for example, can be obtained thus

$$\Delta = \frac{6000}{240} = 25 \text{ mm}$$

$$M - M_{cr} = (\Delta - \Delta_{cr})\frac{M_a - M_{cr}}{\Delta_a - \Delta_{cr}}$$

$$\therefore M = 1.8 + \frac{(25 - 21)(3.48 - 1.8)}{(56.5 - 21)} = 1.99 \text{ kN m}$$

Hence

$$w = \frac{8M}{l^2} = 0.442 \text{ kN/m}$$

This is the method used for evaluating corresponding moments and deflections in the post-buckling range.

The accuracy of deflection determination can be seen from Figure 3.17 which shows a comparison between calculated and experimental values of the centre deflections of a lipped channel beam loaded by pure moment. As can be seen, the calculated deflections are somewhat larger than the experimental ones, thus leading to safe design.

The precise analysis of the behaviour of a cold-formed beam is a very difficult mathematical problem resulting from the nonlinear buckling of the cross-section. The design approach presented in this section simplifies the analysis but at the same time retains the physical characteristics of plate buckling and collapse. The method predicts the deflections of the beam and its maximum load with an accuracy sufficient for engineering purposes. Of course, other design methods have been proposed[5-10] but in all but two references,[9,10] an iterative procedure has been used that results in a more cumbersome set of design calculations.

3.4 SHEAR STRESSES

In addition to keeping direct stresses within defined limits, it is necessary for safe design to ensure that the shear stresses in the beam are not allowed to become dangerously large. Shear stresses occur in beams when the bending moment is varying rapidly, and so tend to be a more important factor in the behaviour of short beams than of long beams. In design two aspects must be taken into account. The first deals with the situation in which, due to the loading and to geometry, the maximum shear stress can reach the material yield condition and localized plastic deformation may take place. The other aspect

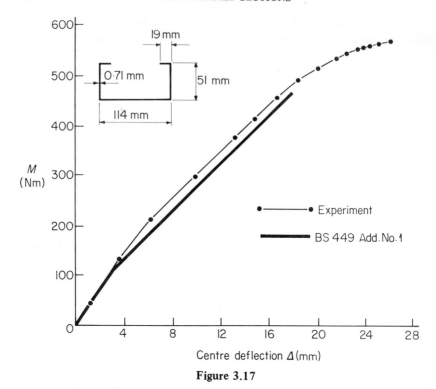

Figure 3.17

concerns instability where, due to the thinness of the sheet from which the section is made, the element subject to the shear will buckle and a consequent loss of strength ensues.

3.4.1 MAXIMUM SHEAR STRESS

The theory for calculating the distribution of shear stress in a member subject to flexure is presented in many textbooks,[1,11] but it may be worth while to review the results of the theory with particular reference to structural members formed from uniform thickness sheet. Also, we will consider only beams having at least one axis of symmetry and loaded normal to that axis.

Consider any general section with this symmetry, Figure 3.18 for example, which is subjected to vertical shear force Q. If we set off a co-ordinate s from one end of the periphery, the shear force per unit length (sometimes called shear flow) at some position a will be

$$q = \frac{QA\bar{y}}{I_x}$$

where $A\bar{y}$ is the first moment of the shaded area about OX. If it is assumed that the shear stress τ is constant through the thickness of the sheet, we have

$$\tau = \frac{QA\bar{y}}{tI_X} \tag{3.9}$$

The application of this formulation can be explained best by means of an example.

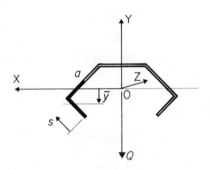

Figure 3.18

Example 3.3
Consider a beam section, Figure 3.19 (a), composed of two cold formed channels welded back to back. It is subjected to a vertical shear force Q having a maximum magnitude of 50 kN. We are required to determine the maximum shear stress. Ignoring the effects of corners, see Figure 3.19 (b), the second moment of area about OX is

$$I_X = 2 \times 100 \times 3 \times (73.5)^2 + \frac{1}{12} \times 6 \times (144)^3 = 4.73 \times 10^6 \text{ mm}^4$$

Taking first the upper flange; the distance[12] from the edge to some position a is s_1, the area of the shaded portion is

$$A = s_1 t_f$$

and the distance y of the centroid of this portion is

$$\bar{y}_1 = \frac{b_w}{2}$$

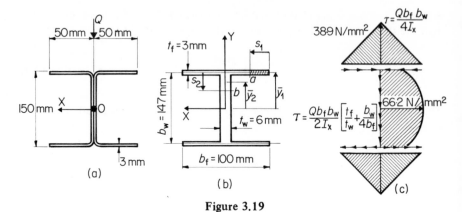

Figure 3.19

Hence the shearing stress is

$$\tau = \frac{Q s_1 t_f b_w}{2 t_f I_x} = \frac{Q s_1 b_w}{2 I_x}$$

$$= \frac{50 \times 10^3 \times 147 \times s_1}{2 \times 4.73 \times 10^6} = 0.78\, s_1 \ \text{N/mm}^2$$

The shear stress along the flange therefore varies from $\tau = 0$ at the edge, to 38.85 N/mm² at the centre where $s_1 = 50$ mm.

In the web at some position b, a distance s_2 from the junction with the flange, the first moment of area comprises the sum of the first moment of area of the flange and the first moment of the portion s_2 of the web.
That is,

$$A\bar{y} = b_f \times t_f \times \frac{b_w}{2} + \left[s_2 \times t_w \times \left(\frac{b_w}{2} - \frac{s_2}{2} \right) \right]$$

and the shear stress is

$$\tau = \frac{Q}{2 I_x} \left[b_f\, b_w\, \frac{t_f}{t_w} + s_2\, (b_w - s_2) \right]$$

This has a parabolic form and varies from

$$\tau = \frac{Q b_f b_w}{2 I_x}\, \frac{t_f}{t_w} = \frac{50 \times 10^3 \times 100 \times 147}{2 \times 4.73 \times 10^6 \times 2} = 38.9 \ \text{N/mm}^2$$

at $s_2 = 0$, i.e. at the junction with the flange, to

$$\tau = \frac{Qb_f b_w}{2I_x}\left[\frac{t_f}{t_w}+\frac{b_w}{4b_f}\right] = 79.2\left[\frac{1}{2}+\frac{147}{400}\right] = 66.2 \text{ N/mm}^2$$

at the middle of the web, i.e. $s_2 = b_w/2$. This distribution of shear stress is shown in Figure 3.19 (c) where also the directions of the shear stresses are indicated. The integrated stresses along the web can easily be shown to be equal to the applied shear force Q. It is, therefore, common practice to define the *average shear stress* as the applied shear force divided by the area of the web; hence for this type of cross-section the maximum and average shear stresses are respectively

$$\tau_{max} = \frac{Qb_f b_w}{8I_x}\cdot\left[2+\frac{b_w}{b_f}\right];\ \tau_{av} \equiv \frac{Q}{b_w t_w}$$

and the ratio of maximum to average shear stress is

$$\frac{\tau_{max}}{\tau_{av}} = \frac{3}{2}\left[\frac{2+b_w/b_f}{3+b_w/b_f}\right] \qquad (3.10)$$

This ratio varies from 1.5 for very small flanges to 1 for very large flanges. Figure 3.20 shows the shear stress distribution resulting in a plain channel section[12] from the application of a force parallel to the web but a distance $b_w^2\ b_f^2 t/4I_x$, from it. It illustrates the necessity for applying the load at this distance if no twisting is to result. This position of load application is known as the *shear centre* and

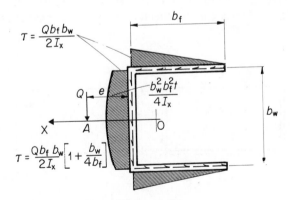

Figure 3.20

we will return to a discussion of this phenomenon in the next chapter. The ratio of the maximum stress to average stress for this section is

$$\frac{\tau_{max}}{\tau_{av}} = \frac{3}{2}\frac{(4 + b_w/b_f)}{(3 + b_w/b_f)}$$

If a section has no axis of symmetry it is of course necessary to take into account the effects of unsymmetrical bending. This is done by calculating the *effective shear forces*, corresponding to the effective bending moments discussed in Section 3.2, and the reader is referred to Reference 11.

Of course the method outlined above must be augmented if there are any openings in the webs. Indeed, in such a circumstance the determination of the maximum shear stress can be difficult as holes tend to be sources of stress concentration. To the shear stresses arising from the varying bending moment must be added those caused by torsion. The next chapter outlines methods for calculating these stresses but it is important to note that the British Specification limits the allowable value of the maximum total shear stress to be equal to $0.46\ \sigma_y$. Since

$$\tau_y = \frac{1}{\sqrt{3}}\sigma_y$$

this provides a margin of 1.25 against any accidental stress concentrations.

3.4.2 WEB BUCKLING DUE TO SHEAR

When the web is thin it is possible for it to buckle when the shear force reaches a critical value. The mechanics of this become clear when we examine the behaviour of a single simply supported plate, Figure 3.21 (a), subjected to a system of shear forces Q. If the plate is flat there will be principal stresses equal to τ, where τ is the applied shear stress. At some critical value of these principal stresses the plate will buckle in the direction of the compressive stress so that it

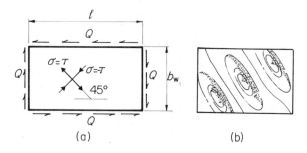

Figure 3.21

forms a 'wash-board' effect with the crests of the ripples at $45°$ to the direction of the shear forces, see Figure 3.21 (b). The critical value of the average shear stress is[13]

$$\tau_{cr} = \left[5.35 + 4\left(\frac{b_w}{l}\right)^2\right] \frac{\pi^2 E}{12(1 - \nu^2)} \left(\frac{t}{b_w}\right)^2 \qquad (3.11)$$

The buckling stress therefore depends mainly on the (b_w/t) ratio, although for short plates it increases rapidly as l/b_w becomes small. For a conservative design formulation the British Standards Specification ignores the effect of length and using $E = 200 \text{ kN/mm}^2$ tabulates numerical values for the buckling stress as

$$\tau_{cr} = \frac{47.0 \times 10^4}{(b_w/t)^2} \qquad (3.12)$$

In equation 3.12 a safety factor of 2 has been incorporated, and since it is independent of σ_y it is applicable to all grades of steel.

Experiments show that, owing to initial lack of flatness, plates subject to shear loads deform from the onset of loading, and that for thin plates the critical stress, equation 3.11, is less than the average stress at which collapse occurs. The mechanism for carrying load in the buckled condition is known as a *tension field* and in fact it is not difficult to see that for a very thin plate most of the load will be supported by the stresses in the tension direction. Collapse occurs by extensive material yielding or by failure of the flanges of the section. A theoretical study of an infinitely long plate[14] (with $\sigma_y = 210 \text{ N/mm}^2$) indicates that the collapse stress (curve A in Figure 3.22) becomes progressively larger than the critical stress (curve

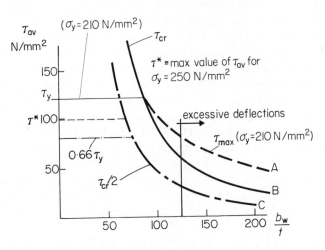

Figure 3.22

B in Figure 3.22) as the slenderness ratio b_w/t of the web increases. However, the effect of imperfections is not included so that the apparently large factor of safety that use of equations 3.12 provides may not be quite so much as it appears in Figure 3.22. For thicker plates the yield stress τ_y will be lower than the critical value and again, to ensure a safe design in the absence of more information regarding the effect of imperfections, the maximum allowable stress is limited to 0.66 α_y.

Since other stresses, such as those direct stresses due to bending, must also be present in the web and contribute to its collapse due to buckling, the above theory in fact considerably simplifies the problem. Moreover, for thin plates, buckle deflections tend to become large, and this is a reason for the slenderness ratio being limited to $b_w/t = 125$ in design calculations. In the use of prefabricated plate for the construction of thin walled beams, such as box girder bridges, this problem is overcome by welding stiffeners onto the web. This is not common practice in the use of cold-formed sections, and for further information on this topic the reader is referred to Reference 13.

3.5 LATERAL BUCKLING

In practice, the majority of cold-formed beams are restrained against lateral deflections, that is deflections perpendicular to the line of action of the loading. The restraint often comes from the floor, roof or walling elements to which the beam is attached. Or, if the beam is part of a framework, some intermediate bracing may be attached to it solely for the purpose of preventing lateral deflections. We have seen in Section 3.1 that if the beam cross-section is asymmetric lateral deflections will tend to increase in magnitude proportionately with the magnitude of the applied load. But the problem we consider in this section concerns beams that may have an axis of symmetry and which are loaded in the plane of that axis. Normally in such conditions lateral deflections will not occur but it has been found in practice that it is possible for the beam to become unstable and for very large lateral deflections to occur at a critical value of the applied load. This type of behaviour is called *lateral buckling.*

The mechanics of this type of buckling are obvious when we examine a very simple example. Suppose we have a beam of rectangular section subjected to a uniform bending moment M about OX as in Figure 3.23. The moment sets up compressive stresses along the top edge of the beam, and when these become sufficiently large, that part of the beam will buckle sideways like a column. Since the lower edge of the beam is subject to longitudinal tensile stresses there is no tendency for that part of the beam to buckle. The beam as a whole therefore deforms primarily by twisting, but also some overall lateral movement takes place.[15]

The main factors affecting lateral instability in a beam are the stiffness of the beam against lateral deflections, the torsional stiffness and the span of the beam.

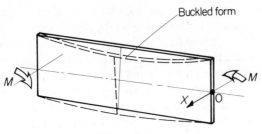

Figure 3.23

Long beams with low lateral stiffness and low torsional stiffness are very prone to buckle laterally. Because of their particular geometry, which gives great flexural rigidity about one axis at the expense of low torsional rigidity and low flexural rigidity about a perpendicular axis, I sections are particularly susceptible to lateral buckling. The elastic buckling of such sections has received a great deal of attention, and lateral buckling under various load conditions has been analysed. For an I section composed of two channel sections fixed back to back, as shown in Figure 3.24, the following expression has been determined[16] theoretically for

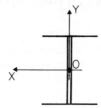

Figure 3.24

evaluation of the critical stress on the compression flange of a beam carrying a uniformly distributed load:

$$\sigma_{cr} = \frac{2.79\,E\,A t^2}{(l/r_Y)^2\,I_X}\left[\sqrt{\left(1 + \frac{l^2}{\pi^2}\frac{J}{\Gamma}\right)} \pm 0.46\right] \qquad (3.13)$$

In this equation, r_Y is the radius of gyration of the section about its weaker axis, i.e. $r_Y{}^2 \equiv I_Y/A$, where A is the cross-sectional area; J and Γ are the uniform torsional and warping stiffnesses, respectively (see Chapter 4), and l is the beam span. The quantity 0.46 within the brackets is preceded by a minus sign if the beam is loaded on the compression flange and by a plus sign if the loading is on the tension flange.

It can be shown that for the vast majority of commercial I sections, whether hot-rolled or cold-formed, the variables in the above expression can be reduced to two, l/r_Y and b_w/t, where b_w/t is the ratio of beam depth to web thickness. On this basis, the elastic buckling stress for cold-formed I sections can be evaluated for various values of b_w/t and l/r_Y. It is found that the values of the theoretical critical stress obtained in this way are low in magnitude only if l/r_Y and b_w/t are very large, i.e. for long, deep beams. For beams of low l/r_Y and b_w/t, the critical stress is generally many times the yield stress for most steels, so that, theoretically, the effects of lateral buckling would appear to be limited to long deep beams.

However, experiments have shown that even short beams of relatively low b_w/t can develop some degree of lateral deflection behaviour in their working range, owing to imperfections in manufacture, etc. These lateral deflections can raise the maximum stress in such a beam and cause premature failure, even though the theoretical elastic lateral buckling load may be much higher. Therefore tendencies to lateral deflection must be taken into account, even for beams with low l/r_Y and b_w/t. The most reliable method found so far to take these deflections into account has been to perform experiments to failure on commercial type beams, and to use the results of these experiments, with a suitable safety factor, in modifying the values obtained from the elastic lateral buckling equation.

The safe working stresses for the compression flanges for I sections found by this semi-experimental, semi-theoretical approach are shown in Figure 3.25 and have been tabulated in the British Standards Specification. These stresses were

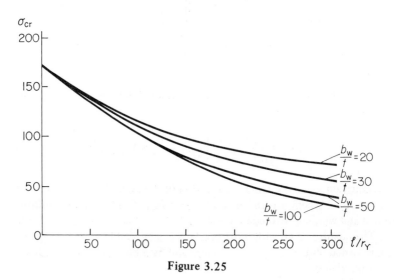

Figure 3.25

originally developed for beams of steel with yield stress 14 ton/in^2 (225 N/mm^2), but later deemed to be suitable, although conservative, for steels with higher yield stress.

3.6 REFERENCES

1 Megson, T. H. G., *Analysis of Thin-Walled Members,* Intertext, 1975.
2 Rhodes, J. and Harvey, J. M., *Proc. ASCE, J. Struct. Div.,* Vol. 97, No. ST6, Aug., p. 2119, 1971.
3 Rhodes, J., 'The non-linear behaviour of thin-walled, beams subjected to pure moment loading, Ph.D. Thesis, University of Strathclyde, 1969.
4 Rhodes, J. and Harvey, J. M., *Aero. Quart.* Vol. XXII Nov. 1971.
5 Bleustein, J. L., and Gjelsvik, A., *Proc. ASCE,* Vol. 96, No. ST7, p. 1535, 1970.
6 Cherry, S., *The Structural Engineer,* Vol. p. 277, 1960.
7 Winter, G., 'Strength of thin steel compression flanges,' Cornell University Engineering Experimental Station, Reprint No. 32, 1947.
8 *Cold Formed Section Specification,* AISI, 1967.
9 Dawson, R. G. and Walker, A. C., *The Structural Engineer,* Vol. 50, p. 95, 1972.
10 Rhodes, J. and Harvey, J. M., Proceedings of IVth International Conference on Experimental Stress Analysis, Cambridge, I. Mech. E. p. 159, 1970.
11 Oden, J. T., *Mechanics of Elastic Structures,* McGraw-Hill, 1967
12 Case, J. and Chilver, A. H., *Strength of Material,* Edward Arnold, 1959.
13 Bulson, P. S., *The Stability of Flat Plates,* Chatto & Windus, 1970.
14 Bergman, S. G. A., *Behaviour of Buckled Rectangular Plates Under the Action of Shearing Forces,* Victor Patersons, Stockholm, 1948.
15 Timoshenko, S. P. and Gere, J. M., *Theory of Elastic Stability,* McGraw-Hill, 1961.
16 Chilver, A. H., 'Structural problems in the use of cold-formed steel sections,' *Proc. ICE,* Vol. 20, p. 233, 1961.

4

Torsion of Thin Walled Beams

4.1 SHEAR CENTRE

It was shown in the previous chapter that when a shear force Q is applied to a thin walled section it will induce a shear flow q around the section. If the shear force is not in a plane of symmetry, as for example in Figure 3.20, the shear stresses tend to cause the beam to twist about its longitudinal axis. The position through which the load must pass if no twisting is to occur is called the *shear centre*. Its position can be calculated from the distribution of shear stresses in the section as is illustrated in the following example.

Example 4.1
The shear stresses in the channel, Figure 3.20, were calculated from equation 3.9 with the assumption that no twisting took place. The torque due to the load is therefore equivalent to that set up as shear forces in the section. Taking moments about A,

$$Q\, e = F\, b_{\mathrm{w}}$$

where e is the distance of the shear centre from the web and F is the force on the flange due to the shear stresses. With

$$F = \frac{1}{2} \times \frac{Q b_{\mathrm{f}} b_{\mathrm{w}}}{2 I_{\mathrm{X}}} \times b_{\mathrm{w}}$$

we have

$$e = \frac{b_{\mathrm{f}}^{2}\, b_{\mathrm{w}}^{2}\, t}{4 I_{\mathrm{X}}} \cong \frac{3 b_{\mathrm{f}}^{2}}{6 b_{\mathrm{f}} + b_{\mathrm{w}}}$$

The position of the shear centre for other cross-sectional shapes is shown in Table 4.1.

94

Table 4.1

	Torsion bending constant Γ	Torsion constant J	e for shear centre	Sectorial coordinate values ω
I-section	$\frac{I_y b_w^2}{4}$ or $\frac{b_w^2 b_f^3 t_f}{24}$	$\frac{2b_f t_f^3 + b_f t_w^3}{3}$ or if $t_w = t_f$ $\frac{t^3}{3}(2b_f + b_w)$	0	$\pm\frac{b_w b_f}{4}$
Channel	$\frac{b_f^3 b_w^2 t}{12} \times \left[\dfrac{2 + \frac{3b_f}{b_w}}{1 + \frac{6b_f}{b_w}}\right]$	$\frac{t^3}{3}(2b_f + b_w)$	$\frac{3b_f^2}{6b_f + b_w}$	$\frac{eb_w}{2}$; $(\frac{b_f-e}{2})b_w$
Lipped channel	$\frac{b_f^2}{6}(4a^3+3b_w^2 a+6b_w a^3)+b_f b_w^2)-I_x e^2$	$\frac{t}{3}(2b_f+2a+b_w)$	$\frac{b_f b_w^2}{I_x}at\left[\frac{1}{2}+\frac{b_f}{4a}-\frac{2}{3}\frac{a^2}{b_w^2}\right]$	$\frac{eb_w}{2}$; $(\frac{b_f-e}{2})b_w$; $-[(\frac{b_f-e}{2})b_w + a(b_f+e)]$; $-\frac{(b_f-e)b_w}{2}$
	$\dfrac{b_f^2[b_w^2(b_f^2+2b_fb_w+4b_fa+6b_wa)+4a^2(3b_fb_w+3b_w^2+4b_fa+2b_wa+a^2)]}{12(2b_f+b_w+2a)}$	$\frac{t^3}{3}(2b_f+2a+b_w)$	0	$\frac{b_w g}{2}$; $+(b_f-g)\frac{b_w}{2}$; $+(b_f-g)\frac{b_w}{2}-ab_f$; $-(b_f-g)\frac{b_w}{2}$ where $g=\frac{b_f^2}{(b_w+2b_f)}$
	$(\frac{b_f^3 b_w^2 t}{12})(\frac{b_f+2b_w}{2b_f+b_w})$	$\frac{t^3}{3}(2b_f+b_w)$	0	$\frac{b_w g}{2}$; $+(b_f-g)\frac{b_w}{2}$ where $g=\frac{b_f^2}{(b_w+2b_f)}$
	$\frac{t^3}{36}(b_f^3+b_w^3)$	$\frac{t^3}{3}(b_w+b_f)$	0	Values of ω for this section are secondary effects only, for the upper and lower surfaces

Practical reasons dictate the plane of load application, and it is generally not through the shear centre. We can account for this in our analysis by superposing two loads, one a shear force through the shear centre, and the other a torque that is proportional to the distance of the load from the shear centre. For example, suppose a shear load is applied at the web of a channel section as shown in Figure 4.1. The stresses due to the force system in Figure 4.1 (a) can be determined as in Section 3.4, and it is the purpose of this chapter to present a method of analysis for evaluating the stresses due to loading systems like that in Figure 4.1 (b). Provided the deformations are small, as they usually are in practice, the stress resultants corresponding to the two loads are additive.

If a cold-formed beam is a component of a frame, the members adjacent to it will apply bending and torsional loading to it. As we will see later, the effect of torsional loading is to induce not only shear stresses but also longitudinal direct

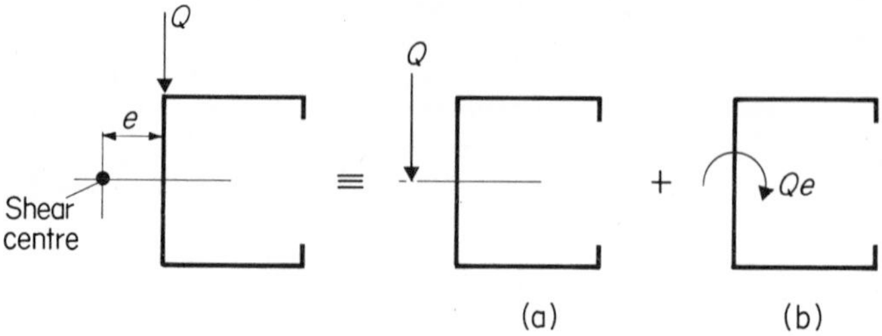

Figure 4.1

stresses. These must be added judiciously to the direct stresses caused by bending so that the maximum stress in the section can be evaluated.

The theory we now develop is based on the assumption that the cross-section remains undistorted during loading and we start by looking at the situation in which no longitudinal restraint is applied to the ends of the beam and a uniform torque acts along the beam axis; this is known as *St. Venant torsion.* Additional stresses are induced if the ends of the beam are restrained in some way or if a varying torque is applied to the beam; this is known as *warping torsion* and is discussed later in this chapter.

4.2 ST. VENANT TORSION

Since most cold-formed sections are composed of flat elements it is reasonable to commence the analysis by looking at the behaviour of a thin strip, Figure 4.2 (a), subjected to a torque T. The shear stresses vary linearly across the thickness t of the strip, Figure 4.2 (b); they are uniform along the length l and are constant across the width b, except near the corners. It can be shown[1] that the stresses τ, torque and geometry are related by

$$\frac{T}{J} = \frac{\tau_x}{2x} = \frac{G\phi}{l} \tag{4.1}$$

where J is a geometric constant called the *torsion constant* and is approximately given by $J = \frac{1}{3} b\, t^3$ for a thin rectangular plate; τ_x is the shear stress at a distance x from the mid-thickness (Figure 4.2 (b)); the twist is ϕ and G is the modulus of shear rigidity.

A section behaves as a collection of plate elements; for example the shear stresses set up in a channel section are shown in Figure 4.3. If we ignore the effect of the stress distribution at the junctions of the plates it can be assumed that each element plate will have a shear stress distribution similar to that in Figure 4.2 (b). Hence, with an accuracy sufficient for engineering purposes, the

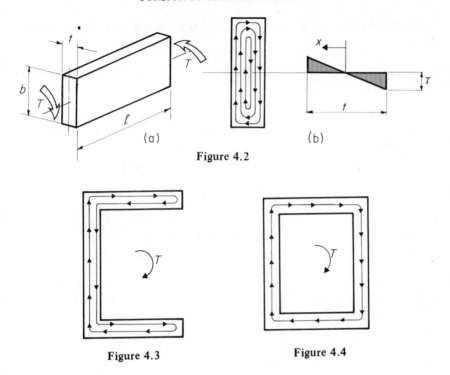

Figure 4.2

Figure 4.3   Figure 4.4

torsion constant for the cross-section is the sum of the torsion constants for each plate, that is

$$J = 2 \times \frac{1}{3} b_1 t^3 + \frac{1}{3} b_2 t^3 = \frac{1}{3} m t^3 \qquad (4.2)$$

where m is the length of the middle line of the cross-section. The maximum stress is on the surface of the sheet, that is $x = t/2$ hence

$$\tau_{max} = \frac{t\,T}{J} = \frac{3T}{mt^2} \qquad (4.3)$$

It is common practice for sections to be cold-formed and then seam welded, thus making closed section tubes. When such a tube is subject to torque loading the shear stress is uniform round the periphery and, within engineering accuracy, can be assumed to be constant across the thickness, see Figure 4.4. The relation between the torque, stress and twist is

$$\frac{T}{J} = \frac{G\phi}{L}; \quad \tau = \frac{T}{2\,At} \qquad (4.4)$$

where A is the area enclosed by the mid-thickness line and the torsion constant is

$$J = \frac{4 A^2 t}{m}$$ (4.5)

Example 4.2

We can evaluate the advantages of using closed tubes to carry torsion loading by considering the section in Figure 4.5, subjected to a torque of 100 Nm. It is made from steel having a shear modulus G of 82.5 kN/mm².

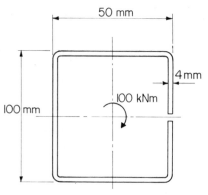

Figure 4.5

Considering it first as an open tube, that is before the welding process, the maximum stress is

$$\tau_{max} = \frac{3T}{mt^2} = \frac{3 \times 100 \times 10^3}{284 \times 16} = 66 \text{ N/mm}^2$$

The torsion constant is

$$J = \frac{1}{3} mt^3 = \frac{1}{3} \times 284 \times 4^3 = 6.06 \times 10^3 \text{ mm}^4$$

so that the rate of twist is

$$\phi/l = \frac{T}{JG} = \frac{100 \times 10^6}{6.06 \times 10^3 \times 82.5 \times 10^3} = 0.20 \text{ rad/m}$$

Considering the section after welding, the stress is

$$\tau = \frac{T}{2At} = \frac{100 \times 10^3}{2 \times 4.42 \times 10^3 \times 4} = 2.82 \text{ N/mm}^2$$

the torsion constant is now

$$J = \frac{4A^2 t}{m} = \frac{4 \times (4.42 \times 10^3)^2 \times 4}{284} = 1.1 \times 10^6 \text{ mm}^4$$

therefore, the rate of twist is

$$\phi/l = \frac{T}{JG} = \frac{100 \times 10^6}{1.1 \times 10^6 \times 82.5 \times 10^3} = 1.1 \times 10^{-3} \text{ rad/m}$$

Thus, by welding the seam and converting the open section to a closed one, we have reduced the maximum stress by over 20 times and considerably increased the torsional stiffness JG.

4.3 WARPING TORSION

When a uniform torque is applied to a member that has no longitudinal restraint applied at its ends, it will twist as indicated in the example section shown in Figure 4.6. But plane sections will not remain plane during the application of

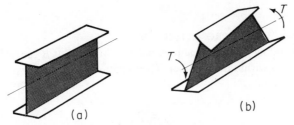

(a)   (b)

Figure 4.6

the torque. Instead, as shown in Figure 4.7 (b), only the web will remain plane while the flanges rotate bodily. The section as a whole therefore departs from its original plane cross-section and this movement is called warping. If now some restraint is applied to the flanges, such as in the cantilever shown in Figure 4.8, the flanges are forced to take up a curvature in the longitudinal direction. This curvature will vary along the beam, and from the appearance of the twisted beam in Figure 4.8 (b) it is as if the warping is restrained by the application of equal but opposite bending moments applied to the flanges on their own plane.

This type of behaviour was extensively investigated by Vlasov[2] who termed this apparent moment system induced in the flanges due to warping restraint *bi-moment,* since it appears as the combination of two moments. The bi-moment has units of force x distance^2, e.g. N mm^2 and we could imagine it to be the combinations of the moments M^*, Figure 4.8 (b), and the depth b_w of the web, that is $B \equiv M^* b_w$. The rigorous theory for the calculation of the bi-moment is

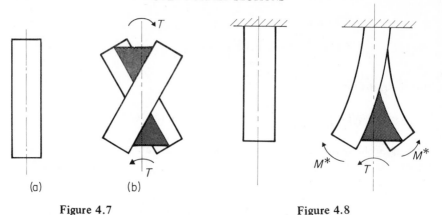

(a)   (b)

Figure 4.7   Figure 4.8

complex, but in this chapter an approximate procedure will be developed which involves an analogy between the bi-moment and the more common treatment of applied moments in the Engineer's bending theory.

4.3.1 CROSS-SECTIONAL PROPERTIES

As with any bending moment, the presence of a bi-moment B causes longitudinal and shear stresses within the member. The longitudinal direct stress is evaluated from the expression

$$\sigma_z = \frac{B\omega}{\Gamma} \tag{4.6}$$

where ω is known as the *sectorial co-ordinate* and is a function of the position on the cross-section at which the stress σ_z is being calculated. The *warping constant* (sometimes called the *torsion-bending constant*) Γ is a function of the cross-sectional dimensions. In analogy, we can compare equation 4.6 to the longitudinal stress σ_z induced by an applied moment M and at a distant y from the neutral axis,

$$\sigma_z = \frac{My}{I} \tag{4.7}$$

The expression given in equation 4.6 for calculating the direct stresses due to torsion can be seen to bear a close resemblance to the corresponding equation giving the stresses arising from a plane bending condition, except that the second moment of area and the distance from the neutral axis to the fibre under consideration are replaced by the warping constant and the sectorial co-ordinate, respectively. The warping constant represents the resistance to twisting provided by the walls of the section, and details of its evaluation are given by Timoshenko.[3] The procedure is tedious if followed from basic principles, and values of the warping constant for various common cross-sectional shapes are shown in Table 4.1.

Unlike the warping constant, the sectorial co-ordinate does not have a fixed value for a particular cross-sectional shape, but varies around the section. It is calculated for any point on the cross-section from the equation

$$\omega = \int_0^s h \ ds$$

where h is the perpendicular distance from the shear centre to a tangent at the point under consideration, and s is the distance from some chosen origin to the same point measured along the middle line of the section. To illustrate the calculation of the sectorial co-ordinate, consider the channel section shown in Figure 4.9. Point A is the shear centre, D is the point where the axis of symmetry intersects the web, O is the origin and the direction of viewing is in the positive

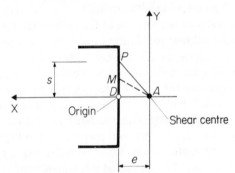

'Z' axis positive into the plane of the diagram

Figure 4.9

direction of the Z axis. The value of ω at any point P on the web is given by the product $e \ s$, and is equal to twice the area swept out by a radius vector AM rotating from AD to AP. The sign of the sectorial co-ordinate is positive if the radius vector rotates clockwise to reach the point P.

A diagrammatic representation of the sectorial co-ordinate at any position in a channel section is shown in Figure 4.10. Since the radius vector is rotating clockwise, the values for the points on the web above the AX axis are positive. For the flanges, the sectorial co-ordinate decreases as the distance from the web increases. At point C, which is the same distance from the web as the shear

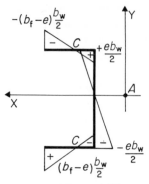

Figure 4.10

centre A is from D, the sectorial co-ordinate is zero. Thereafter the value is of
opposite sign and increases linearly to a maximum at the free edge of the flange.

Now let us consider the same channel section, but this time placing the
section as shown in Figure 4.11 (a). We are still viewing the section in the
positive Z direction, but now the rotation of the radius vector AM in moving
from AD to AP is anticlockwise so that the signs of the sectorial co-ordinates
for portions of the web above the AX axis are negative. By completing the
sectorial co-ordinate diagram in the same way as before, the values of the
sectorial co-ordinates around the complete section may be established as those
shown in Figure 4.11 (b). It will be seen that the sectorial co-ordinate values are
of opposite sign to those previously obtained. Since equation 4.6 has shown
that the longitudinal stress due to torsion is proportional to the value of the
sectorial co-ordinate, the implication is that a reversal in the sign of the sectorial
co-ordinate produces a reversal in the sign of this longitudinal torsional stress.
If the loading on the section shown in Figure 4.11 (a) were such that any
rotation of that section was in the same direction as any rotation of the section
in Figure 4.9 (e.g. both clockwise), then there would indeed be a reversal of
the stresses. However, if we consider a practical loading case, such as a load
applied over the full width of the flange, then the load would be on the opposite
side of the shear centre in Figure 4.9 compared to Figure 4.11 (a), and the twist
of the section in Figure 4.11 (a) would be opposite to that in Figure 4.9 (see
Figure 4.1), thus leading to a reversal of the sign of the bi-moment, and making
the longitudinal torsional stresses in the two systems equal in sign.

A similar effect may be demonstrated if the direction of viewing is negative,
showing that it is extremely important to consider the sign of the bi-moment on
the section as well as the sign of the sectorial co-ordinate.

Values for the sectorial co-ordinates for various cross-sectional shapes are
given in Table 4.1, all are viewed in the positive direction of the Z axis.

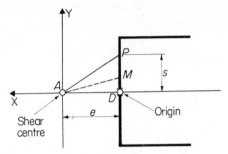

Z axis positive into the plane of the diagram

(a)

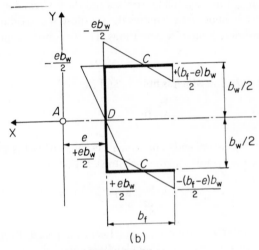

(b)

Figure 4.11

4.3.2 CALCULATION OF THE WARPING STRESSES

The bi-moment which arises from the bending of the flanges is proportional to the curvature of these flanges in their own plane. This curvature is related to the twist ϕ of the section so that

$$B = -E\Gamma \frac{d^2\phi}{dz^2} \tag{4.8}$$

Also, the torque T_ω that can be attributed to the warping–bending of the flanges is equal to the rate of change of the bi-moment, that is

$$T_\omega = \frac{dB}{dz} \tag{4.9}$$

The total torque T on a section is the sum of the torque T_s due to the St Venant shear stresses and that due to warping–bending. Hence

$$T = T_s + T_\omega$$

but

$$T_s = GJ \frac{d\phi}{dz} \quad \text{(see equation 4.1)}$$

therefore from equations 4.8 and 4.9,

$$-E\Gamma \frac{d^3\phi}{dz^3} + GJ \frac{d\phi}{dz} = T \tag{4.10}$$

The evaluation of the longitudinal stresses and shear stress due to the application of a torque to a member now requires the solution of equation 4.10 together with the appropriate boundary conditions. The two most common forms of end supports are:

(1) *Fixed end*—one which is built-in and can neither twist nor warp, that is

$$\phi = 0; \frac{d\phi}{dz} = 0 \tag{4.11}$$

(2) *Simply supported end*—one which cannot twist but is free to warp and is therefore free of longitudinal stresses, that is

$$\phi = 0; \frac{d^2\phi}{dz^2} = 0 = B \tag{4.12}$$

Later in this chapter we shall discuss practical examples of these types of support.

It is not possible to provide a completely general solution to equation 4.10 that can be applied to all structures. However, the general method of solution is identical for all cases and in order to demonstrate the technique two cases will be examined; a single span beam and a continuous three span beam.

First we shall consider a single span beam which is subjected to a concentrated torque at the centre of the span l. The end conditions are simply supported. Since the beam and loading are symmetrical, the solution need be found for half the beam only, that is in the range $0 \leqslant z \leqslant l/2$, where the origin is taken at a support. The problem is solved by obtaining the general solution of the torsion equation 4.10 and then by substituting the boundary conditions into this result; thus we obtain a result that is specific to this particular case. Writing equation 4.10 in its more common form,

$$E\Gamma \frac{d^4\phi}{dz^4} - GJ \frac{d^2\phi}{dz^2} = m; \left(m \equiv -\frac{dT}{dz}\right) \tag{4.13}$$

we can use equation 4.8 to rewrite equation 4.13

$$\frac{d^2 B}{dz^2} - \lambda^2 B = -m; \lambda \equiv \sqrt{\left(\frac{GJ}{E\Gamma}\right)} \qquad (4.14)$$

The general solution of this equation takes the form

$$B = A \cosh \lambda z + C \sinh \lambda z + B_0 \qquad (4.15)$$

where A and C are constants which can be determined for any specific case from the relevant boundary and loading conditions, and B_0 is the particular solution which is dependent upon the load application. For a beam subjected to concentrated torques, the rate of change of torque between the positions of load application is zero. Hence $m = 0$ in equation 4.14, giving $B_0 = 0$ in equation 4.15. Therefore,

$$B = A \cosh \lambda z + C \sinh \lambda z \qquad (4.16)$$

For the case considered here the boundary conditions are

$$B = 0 \text{ at } z = 0 \quad \text{(see equation 4.12)} \qquad (4.17)$$

Also, from symmetry $d\phi/dz = 0$ at $z = l/2$ hence for the portion of the beam $0 \leqslant z \leqslant l/2$ we have from equations 4.13 and 4.8

$$\frac{dB}{dz} = \frac{T}{2} \text{ at } z = \frac{l}{2} \qquad (4.18)$$

From equations 4.16 and 4.17 we have

$$0 = A \cosh(0) + C \sinh(0) \qquad (4.19)$$

and from equations 4.16 and 4.18

$$\frac{T}{2} = \lambda A \sinh \frac{\lambda l}{2} + \lambda C \cosh \frac{\lambda l}{2} \qquad (4.20)$$

Hence, $A = 0$ and

$$C = \frac{T}{2\lambda \cosh \lambda l/2} .$$

Substituting these values of A and C in equation 4.16 now gives the solution for B for the problem under examination as

$$B = \frac{T \sinh \lambda z}{2\lambda \cosh \lambda l/2}, 0 \leqslant z \leqslant \frac{l}{2}; \lambda \equiv \sqrt{\left(\frac{GJ}{E\Gamma}\right)} \qquad (4.21)$$

and the bi-moment at mid-span is

$$B = \frac{T}{2\lambda} \tanh \frac{\lambda l}{2} \text{ at } z = \frac{l}{2} \qquad (4.22)$$

Using other boundary and loading conditions together with equation 4.15 we can generate corresponding solutions for a wide variety of practical cases of structural members loaded in torsion. Unfortunately, the expressions describing the variation of bi-moment, such as equation 4.22, are frequently intricate and not suitable for use in design analysis. But usually we are concerned only with knowing the maximum value of the bi-moment and this fact enables us to develop a simpler presentation restricted to the variation of the maximum bi-moment with applied torque.

Suppose that the beam section under investigation is such that the component of torque $GJ\,d\phi/dz$ due to St. Venant shear stresses is much smaller than that due to warping–bending $E\Gamma\,d^3\phi/dz^3$. We can therefore without much loss of accuracy neglect the former term in equation 4.10; hence the governing equation is now,

$$-E\Gamma\frac{d^3\phi}{dz^3} = T$$

or,

$$E\Gamma\frac{d^4\phi}{dz^4} = m \tag{4.23}$$

This is very similar in form to the equation of plane bending theory relating the deflection Δ to the intensity of the applied lateral load w.

$$El\frac{d^4\Delta}{dz^4} = w \tag{4.24}$$

Consequently, since we have similar boundary conditions in plane bending as in warping torsion, it follows that solutions can be obtained for equation 4.23 in a manner similar to equation 4.24 for plane bending.

For example, taking the simply supported beam considered above, we had the torque applied at the centre and therefore $m = 0$. Writing equation 4.23 as

$$E\Gamma\frac{d^2B}{dz^2} = -m \tag{4.25}$$

we have the solution

$$B = Az + C$$

Using the boundary conditions of equations 4.17 and 4.18 results in the solution $B = (T/2)\,z$ and hence

$$B = \frac{Tl}{4} \text{ at } z = l/2 \tag{4.26}$$

The corresponding values of the bi-moment for a variety of boundary conditions and types of loading are shown in Table 4.2 (positive torques as shown in the table are clockwise when viewed in the positive z direction). But obviously we cannot just ignore the contribution from the St. Venant shear stresses since their effect may well be considerable. We take these into account by classifying the values of the bi-moment obtained from equation 4.25 as the *approximate bi-moment* B_{app}

Table 4.2

Load condition	Approximate bimoment (B_{app}) At ends of beam	At load application point	Approx. angle of twist (rotation) at load application point
	0	$-\dfrac{T\ell}{4}$	$\dfrac{T\ell^3}{48E\Gamma}$
	0	$-T\alpha\ell(1-\alpha)$	$\dfrac{T\alpha^2(1-\alpha)^2\ell^3}{3E\Gamma}$
	0	$-\dfrac{m\ell^2}{8}$ at centre $(z=\ell/2)$	$\dfrac{5m\ell^4}{384E\Gamma}$ at centre $(z=\ell/2)$
	At A $B_{app}=+T\alpha\ell(1-\alpha)^2$ At C $B_{app}=+T\alpha^2\ell(1-\alpha)$ $+\dfrac{mL^2}{12}$	$-2T\alpha^2\ell(1-\alpha)^2$ $-\dfrac{m\ell^2}{24}$ at centre	$\dfrac{T\alpha^3(1-\alpha)^3\ell^3}{3E\Gamma}$ $\dfrac{m\ell^4}{384E\Gamma}$ at centre $(z=\ell/2)$

and by comparing the approximate values with the solution B obtained from the more accurate formulation, equation 4.14, we can evaluate the error. Equation 4.26 indicates that the value at mid-span calculated by the approximate method is constant for all sections. However, examination of the accurate solution as given in equation 4.22 shows that the bi-moment is dependent on the parameter λl, termed the *beam parameter,* and the errors arising from the use of the approximate method must therefore be calculated for various values of the beam parameter. The amount of error can be evaluated by considering the ratio of the approximate to accurate values of bi-moment,

$$\frac{B_{app}}{B} = \frac{Tl \times 2\lambda \cosh \lambda l/2}{4 \times T \times \sinh \lambda l/2} \tag{4.27}$$

The results of this calculation are shown in Figure 4.12 for various values of λl, from which it may be seen that for low values the error is small but increases as λl also increases. Thus, although the approximate method is seen to give acceptable results for $\lambda\!\ell \gtrsim 1$, for greater values of $\lambda\!\ell$ the approximate solution overestimates the bi-moment value, leading to over-conservative design.

In order to maintain the simplicity of the approximate method, and at the same time to achieve the overall economy of design that the accurate method would permit, Figure 4.13 has been produced. This shows a graph of correction factor F against beam parameter $\lambda\!\ell$, where $B = F \times B_{app}$, and

$$F = \frac{2 \sinh \lambda\!\ell/2}{\lambda L \cosh \lambda\!\ell/2} \tag{4.28}$$

Thus accurate values of bi-moment can be readily calculated from knowledge of the approximate result and the correction factor. Figure 4.13 is applicable only

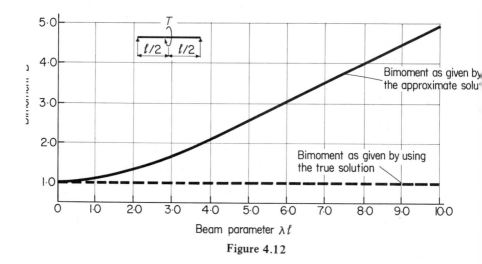

Figure 4.12

to the case examined above, that is a simply supported beam subjected to a concentrated torque, but of course the basic approach may be applied to other cases. Values of the approximate bi-moment for a variety of common loading and support conditions are given in Table 4.2, and the corresponding graphs of correction factors are shown in Appendix 1, p. 159, Figures A1.1–A1.6.

Example 4.3
In this example we shall calculate the longitudinal stresses in a channel beam, see Figure 4.14 (a), subjected to a uniformly distributed positive torque of

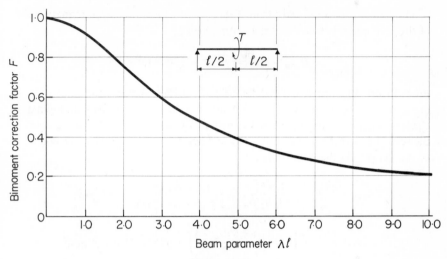

Figure 4.13

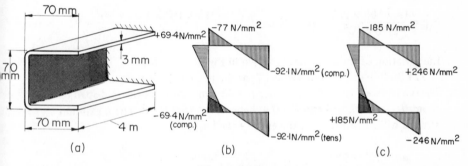

Figure 4.14

60 Nm/m when its ends are: (1) simply supported and (2) built-in (E = 200 kN/mm², G = 82.5 kN/mm²).

From Table 4.1,

$$J = \frac{1}{3} mt^3 = \frac{1}{3} \times (2 \times 68.5 + 67) \times 27 = 1.84 \times 10^3 \text{ mm}^4$$

$$\Gamma = \frac{b_f^3 \, b_w^2 \, t}{12} \left[\frac{2 + 3 \, b_f/b_w}{1 + 6 \, b_f/b_w} \right] = \frac{68.5^3 \times 67^2 \times 3}{12} \left[\frac{2 + 3 \times 68.5/67}{1 + 6 \times 68.5/67} \right] = 2.56 \times 10^8 \text{ mm}^6$$

$$\lambda l \equiv l \times \sqrt{\left(\frac{GJ}{EI} \right)} = 4 \times 10^3 \times \left[\frac{82.5 \times 1.84 \times 10^3}{200 \times 2.56 \times 10^8} \right]^{\frac{1}{2}} = 6.89$$

(1) Simply supported ends; from Table 4.2, the maximum approximate value of the bi-moment is $-mL^2/8$ at the centre, hence

$$B_{app} = -\frac{60 \times 16}{8} = -120 \text{ N m}^2$$

From Figure A1.2 the correction factor is $F = 0.15$ so the accurate value of bi-moment is $B = -120 \times 0.15 = -18$ N m^2. The magnitude of the longitudinal stress is evaluated using equation 4.6, $\sigma_z = B\omega/\Gamma$, and in Table 4.1 we see that the sectorial co-ordinate varies linearly and has its greatest values at the edges of the flanges. At these positions,

$$\omega_1 = \frac{e\, b_w}{2} = \frac{1 \cdot 5\, b_f^2}{1 + 6\, b_f/b_w} = \frac{1.5 \times 68.5^2}{1 + (6 \times 68.5/67)} = 987 \text{ mm}^2$$

$$\omega_2 = \left(\frac{b_f - e}{2}\right) b_w = \left(\frac{68.5 - 29.4}{2}\right) \times 67 = 1.31 \times 10^3 \text{ mm}^2$$

Thus the stresses are

$$\sigma_{z1} = \frac{18 \times 10^6 \times 987}{2.56 \times 10^8} = 69.4 \text{ N/mm}^2\,;\, \sigma_{z2} = \frac{18 \times 10^6 \times 1.31 \times 10^3}{2.56 \times 10^8} = 92.1 \text{ N/mm}^2$$

The variation of these stresses is shown in Figure 4.14 (b).

(2) Built-in ends; from Table 4.2, at the beam supports, $B_{app} = mL^2/12$ and $F = 0.615$ (see Figure A1.6), hence $B = 48$ Nm^2. Now, at the mid-span of the beam $B_{app} = -mL^2/24$ and $F = 0.4$, hence $B = -16.3$ N m^2. The maximum stresses will therefore be at the ends of the beam.

$$\sigma_{z1} = \frac{48 \times 10^6 \times 987}{2.56 \times 10^6} = 185 \text{ N/mm}^2\,;\, \sigma_{z2} = \frac{48 \times 10^6 \times 1.31 \times 10^3}{2.56 \times 10^8} = 246 \text{ N/mm}^2$$

4.3.3 CALCULATION OF ANGLE OF TWIST

The effect of restraining the ends from warping, in addition to the restraint against twisting, is therefore to increase the longitudinal stresses, as seen in Figure 4.14 (c). This disadvantage is offset by an increase in the stiffness of the beam, that is the twist is much less for a built-in beam than it is for a corresponding beam with simply supported ends.

In this example for case (1) the approximate angle of twist, or rotation, ϕ_{app} is maximum at mid-span, from Table 4.2

$$\phi_{app} = \frac{5\, ml^4}{384\, E\Gamma}$$

Figure 4.15 is a graph of the rotation correction factor F^* for a range of values of λl. For this beam, $\lambda l = 6.89$, so that $F^* = 0.17$ and we can calculate the accurate value of the maximum angle of rotation ϕ_{max} as

$$\phi_{max} = F^* \; \phi_{app} = 0.17 \; \frac{5 \; ml^4}{384 \; E\Gamma}$$

$$= \frac{0.17 \times 5 \times 60 \times (4000)^4}{384 \times 200 \times 10^3 \times 2.56 \times 10^8} = 0.66 \; \text{rad}$$

Expressions for the maximum angle of rotation for beams having a variety of boundary and loading conditions are given in Table 4.2 (Notice the similarity to the expressions for the maximum deflection of beams subject to plane bending.) The graphs of the corresponding correction factors are shown in Appendix 1, p. 163, Figures A1.7–A1.10.

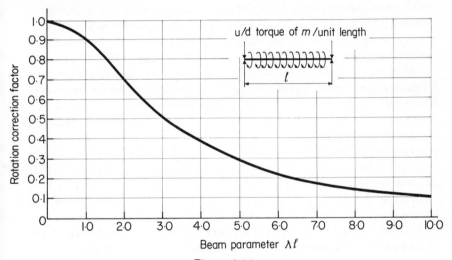

Figure 4.15

Continuing with the example; the maximum value of the rotation for case (2) is given approximately (Table 4.2) as

$$\phi_{app} = \frac{ml^4}{384 \; E\Gamma}$$

and from Figure A1.10 we have the correction factor as $F^* = 0.45$, hence

$$\phi_{max} = 0.45 \; \frac{ml^4}{384 \; E\Gamma} = 0.35 \; \text{rad}$$

which is about half the value for the simply supported beam.

Since the longitudinal warping stresses at any point are proportional to the value of the sectorial co-ordinate at that point, the distribution of warping stress around the section may be represented by the sectorial co-ordinate diagram (Table 4.1). From this it is obvious that considerable variation of the warping stresses takes place around the section, so that if the permissible stress is limited to a maximum of $0.65 \sigma_y$ (as with bending stresses), much of the section is grossly understressed and an uneconomic design results. Addendum 1 to BS 449 allows for this by permitting the maximum stress due to combined bending and torsion to be σ_y, the argument being that this intensity of stress will only occur in an extremely localized portion of the section, and that the section will not therefore be generally overstressed.

4.3.4 PRACTICAL BOUNDARY CONDITIONS

Generally when a beam is loaded torsionally and warping is somehow restrained, the sectorial co-ordinates (Table 4.1) dictate the distribution of the resulting longitudinal stresses. But these stresses can only develop at the ends if the corresponding displacements are effectively prevented. In other words if the section is attached to another member only at those points where the sectorial co-ordinates, and therefore the warping stresses, are zero, the beam will be effectively unrestrained against warping. For example, consider the I section in Figure 4.16 (a): the sectorial co-ordinate distribution is shown in Figure 4.16 (b) where we note that it is zero along the web. Therefore if the beam is attached to another member by a cleat bolted solely at the web, only twist restraint will be provided and no warping stresses will be induced. Alternatively, if the flanges are firmly attached to the adjacent member, as for example in Figure 4.16 (d), not only is rotational restraint provided but also a large degree of warping restraint. In this situation the longitudinal stresses may be considerably increased but the amount of twist deformation will be reduced.

In the case of a Z section, Figure 4.17 (a), it is difficult to avoid some degree of warping restraint since the sectorial co-ordinate is zero only at two points on the cross-section, see Figure 4.17 (b). To achieve solely rotational restraint it would be necessary to use two very narrow angle pieces as shown in Figure 4.17 (c). The use of cleats attached to the web, Figure 4.17 (d), implies rotational restraint together with some warping restraint but avoids the high stresses that would exist if the flanges were attached to the adjoining member.

In all situations where the joint is fully welded it must be realized that complete warping restraint is involved and the weld must be sufficiently strong to withstand the warping stresses.

4.3.5 TWISTING OF THIN WALLED TIES

Of course if twisting induces longitudinal stresses in situations of warping restraint, it follows that longitudinal loads can induce twisting. Thus, if a longi-

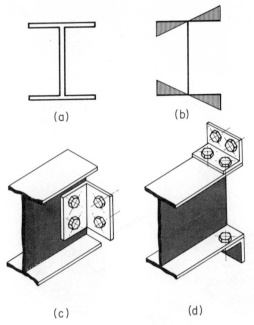

(a) (b)

(c) (d)

Figure 4.16

tudinal load P, Figure 4.18, is applied to Z section at a position where the sectorial co-ordinate is not zero, for example in the web, the member will twist. In this case it can be shown[4] that due to the action of the tensile load the member twists an amount

$$\phi = \frac{3P}{G} \left(1 - \text{sech } \lambda l\right) \frac{b_f^2 \, b_w}{(b_w + 2b_f)^2 \, t^3} \text{ rad}$$

When the member is very long the angle of twist approximates to

$$\phi \simeq \frac{3p}{G} \frac{b_f^2 \, b_w}{(b_w + 2b_f)^2 \, t^3}$$

Notice that this twist is inversely proportional to t^3 so that twisting of tension members can be quite important for cold-formed sections. In general twisting will take place in thin walled ties and is only avoided if the loads are introduced at points on the cross-section where the sectorial co-ordinate is zero or if the tensile stress is uniform over the whole cross-section.

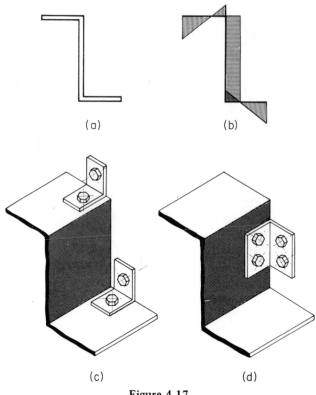

(a)   (b)

(c)   (d)

Figure 4.17

4.4 THREE SPAN CONTINUOUS BEAM

4.4.1 FORMAL SOLUTION

Now we consider a three span continuous system exemplified by the channel section shown in Figure 4.19. The supports are such that they can be considered to form simple supports with respect to the bending load w, and give restraint against twisting but not warping with respect to the torque loading T. The bending and torque loadings are both taken relative to the shear centre line. Any load not acting at the shear centre line can be transformed into the system analysed here.

Stress distributions due to the loading w can be obtained from the plane bending theory outlined in Chapter 3 and do not require further discussion here. The longitudinal stresses resulting from torsion and the induced warping restraint

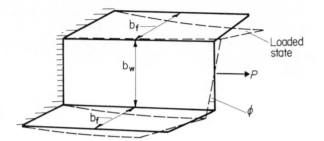

Figure 4.18

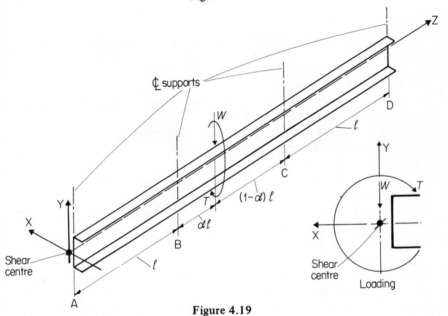

Figure 4.19

can be calculated from the basic torsion equation 4.13 in which for this case, since we have a 'point' torque, $m = 0$,

$$E\Gamma \frac{d^4\phi}{dz^4} - GC \frac{d^2\phi}{dz^2} = 0 \tag{4.29}$$

which has a general solution

$$\phi = A \cosh \lambda z + C \sinh \lambda z + Dz + H \tag{4.30}$$

The example considered here is made more general by letting the middle span have elastic and section properties different from those of the outer spans, and by applying the load in the middle span at some position distance αl $(0 \leqslant \alpha \leqslant 1)$ from an internal support.

There are consequently four typical sections of the beam, the two outer spans and the two portions of the middle span into which it is divided by the load. For each section we have a general solution of the type given in equation 4.30.

$$\phi_i = A_i \cosh \lambda_i z_i + C_i \sinh \lambda_i z_i + D_i z_i + H_i \qquad (4.31)$$

where

$$\lambda_i \ (i = 1 \text{ and } 4) = \sqrt{\left(\frac{G_a J_a}{E_a \Gamma_a}\right)} \ ; \ \lambda_i \ (i = 2 \text{ and } 3) = \sqrt{\left(\frac{G_b J_b}{E_b \Gamma_b}\right)}$$

The problem must be divided into four sections since, because there is no symmetry of loading, equation 4.30 is not continuous over the centre span, but becomes discontinuous at the load application point. Similarly, the absence of symmetry resolves the end spans into separate problems. Hence we have four equations (one for each part of the structure) involving a total of 16 coefficients to be evaluated. In order to perform a solution to these equations, 16 known conditions on the angle of twist ϕ and its derivatives must now be formulated. This is achieved in a similar manner to the method used in the solution of the single span beam discussed earlier, that is by considering ϕ and its derivatives at points of support and load application. These may be conditions of compatibility or equilibrium between one portion of the beam and another.

The resulting set of simultaneous algebraic equations can either be solved formally to give explicit expressions for the coefficients A_i etc., or the problem can be programmed for a computer and the equations solved using matrix algebra for particular values of the geometry and positions of load application. Certainly if we have a structure with more than three members, this latter approach is the only practicable one, and the reader is advised to read Reference 5 for further details.

Nevertheless, formal solutions are valuable as comparisons to the approximate solutions obtained by numerical methods using computers. The formal solution for this three span beam is provided in Appendix 2, p. 165, and it is not difficult to realize that this type of solution is impracticable for use as the basis of hand calculations in a design office. However, the task of programming the formal solution for a small computer is not too difficult, and it then provides an accurate design formulation for the three span beam. We will go on now to develop a simpler, but more approximate, method whereby the torsion stresses in continuous thin walled beams can be calculated.

4.4.2 BI-MOMENT DISTRIBUTION TECHNIQUE

We saw earlier that by neglecting the St. Venant torsion term, the torsion equation reduces to

$$E\Gamma \frac{d^2 \phi}{dz^2} = -B \qquad (4.32)$$

Further, since the stresses due to a bi-moment B can be obtained from $\sigma_z = B\omega/\Gamma$, it is evident that for design purposes the problem becomes one of determining the distribution of bi-moment in the various spans of the beam. Equation 4.32 has a form similar to that for plane bending, and as we have well established methods for determining the distribution of bending moments in redundant structures it seems worth while considering whether these approaches could be used for the bi-moment problem. In this section, therefore, we shall analyse the torsion of the three span beam using a technique of *bi-moment distribution* directly analogous to the Hardy-Cross moment distribution method so commonly used for plane bending situations. In presenting this technique a particular beam is considered, details of which are shown in Figure 4.20.

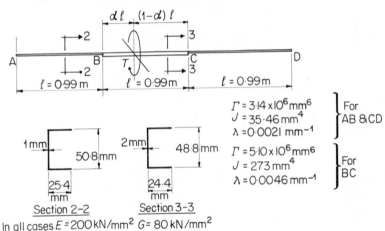

$\Gamma = 3.14 \times 10^6 \, \text{mm}^6$
$J = 35.46 \, \text{mm}^4$
$\lambda = 0.0021 \, \text{mm}^{-1}$ For AB & CD

$\Gamma = 5.10 \times 10^6 \, \text{mm}^6$
$J = 273 \, \text{mm}^4$
$\lambda = 0.0046 \, \text{mm}^{-1}$ For BC

Section 2-2   Section 3-3

In all cases $E = 200 \, \text{kN/mm}^2$ $G = 80 \, \text{kN/mm}^2$

Figure 4.20

Distribution factor

Following directly the moment distribution method, the distribution factors D at the simple support A can be calculated as

$$D_{BA} = \frac{3/4 \, (E\Gamma_{BA}/l_{BA})}{(3/4 \, E\Gamma_{BA}/l_{BA} + E\Gamma_{BC}/l_{BC})} = 0.315$$

and, since $D_{BC} = (1 - D_{BA})$, we have $D_{BC} = 0.685$. By symmetry, $D_{CD} = 0.315$ and $D_{CB} = 0.685$.

Fixed end bi-moments

Considering unit torque and span ($T = l = 1$) the fixed end bi-moments for the unit torque applied at some distance from the left hand support are (see Table 4.2)

$$B_{BC}^{F} = (\alpha - 2\alpha^2 + \alpha^3) \equiv -K$$

$$B_{CD}^{F} = \alpha^3 - \alpha^2 \equiv L$$

Carry-over factor

Again in direct analogy with plane bending the carry-over factor is taken as 0.5.

The calculation to determine the bi-moment by the relaxation technique is shown in Table 4.3, in which we see that the final results are

$$B_{BA} = -0.356\,\alpha + 0.591\alpha^2 - 0.235\alpha^3$$

$$B_{CD} = +0.121\alpha + 0.114\alpha^2 - 0.235\alpha^3$$

Table 4.3

Fixed end bi-moments

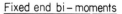

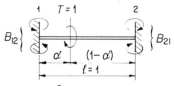

$$B_{12} = \alpha(1-\alpha)^2 = +(\alpha - 2\alpha^2 + \alpha^3) \equiv -K$$

$$B_{21} = -\alpha^2(1-\alpha) = -(\alpha^2 - \alpha^3) \equiv +L$$

Bi-moment distribution

	·315	·685		·685	·315	
A	B	$-K$		$+L$	C	D
	$+·315\,K$	$+685K$		$-·685\,L$	$-·315\,L$	
		$-·343\,L$		$+·343\,K$		
	$+·108\,L$	$+·235\,L$		$-·235\,K$	$-·108\,K$	
		$-·118\,K$		$+·118\,L$		
	$+·037\,K$	$+·081\,K$		$-·081\,L$	$-·037\,L$	
		$-·041\,L$		$+·041\,K$		
	$+·013\,L$	$-·028\,L$		$-·028\,K$	$-·013\,K$	
		$-·014\,K$		$+·014\,L$		
	$+·004\,K$	$+·010\,K$		$-·010\,L$	$-·004\,L$	
	$+C$	$-C$		$+D$	$-D$	

Where $C = 0·356\,K + 0·121\,L$

  &$D = 0·121\,K + 0·356\,L$

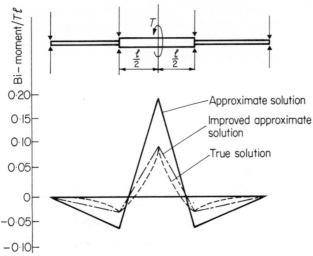

Figure 4.21

These values obtained from this approximate analysis are compared in Figure 4.21 with the corresponding results from the more accurate formal solution.

Obviously, from this result we can conclude that, although the calculation of the bi-moment has been made more practicable, the elimination of the St. Venant torsion term from equation 4.29 has oversimplified the problem and introduced unacceptable errors. But there is a compromise by which the above technique can be used to determine more accurate values of the bi-moment and yet retain the essential simplicity of the moment distribution method. This is achieved by including the *GJ* term in the torsion equation and calculating accurate values for the distribution factors and carry-over factor.

To facilitate this more accurate approach to the analysis, the relevant case of a two span continuous beam, Figure 4.22, built-in at one end (complete

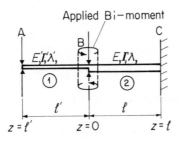

Figure 4.22

warping restraint), simply supported at the other end (no warping restraint), and continuous over an internal support has been analysed[5] using the formal approach presented in equations A2.1–A2.9. From that analysis we have the distribution factors at the internal support as

$$D_{BA} = \frac{-E_2\Gamma_2\,\lambda_2^2 l_2\,(2 - 2\,\cosh\lambda_1 l_1 + \lambda_1 l_1\,\sinh\lambda_1 l_1)}{[E\Gamma_1^2\,(\lambda_2 l_2\,\coth\lambda_2 l_2)\,(\sinh\lambda_1 l_1 - l\,\cosh\lambda_1 l_1)}$$
$$- E_2\Gamma_2\lambda_2^2\,(2 - 2\,\cosh\lambda_1 l_1 + \lambda_1 l_1\,\sinh\lambda_1 l_1)]$$

$$(4.33)$$

$$D_{BC} = \frac{E_1\Gamma_1^2\,(\lambda_2 l_2\,\coth\lambda_2 l_2 - 1)\,(\sinh\lambda_1 l_1 - \lambda_1 l_1\,\cosh\lambda_1 l_1)}{[E\Gamma_1^2\,(\lambda_2 l_2\,\coth\lambda_2 l_2 - 1)\,(\sinh\lambda_1 l_1 - \lambda_1 l_1\,\cosh\lambda_1 l_1)}$$
$$- E_2\Gamma_2\lambda_2^2 l_1\,(2 - 2\,\cosh\lambda_1 l_1 + \lambda_1 l_1\,\sinh\lambda_1 l_1]$$

$$(4.34)$$

The accurate value for the bi-moment carry-over factor was also determined and found to be

$$\frac{\sinh\lambda_1 l_1 - \lambda_1 l_1}{\sinh\lambda_1 l_1 - \lambda_1 l_1\,\cosh\lambda_1 l_1}$$

$$(4.35)$$

When the accurate distribution factors and carry-over factors, as given by equations 4.33–4.35, are used in the bi-moment distribution technique presented above, the resulting bi-moment distribution is very close to the accurate solution. This is shown in Figure 4.21 where we compare the accurate solution with the approximate solution and this improved approximate solution.

Although the results obtained by using this technique are more accurate than those obtained by the approximate method, both the analysis and the results are such that the method is still not suited for hand computation. This may be readily appreciated from examination of equations 4.33–4.35. The technique does have the great advantage over the formal solution, however, in that, once the distribution and carry-over factors for the problem under examination are known, the method may be easily used in design analyses. But a designer will not want to go through a complicated procedure to obtain values of the various factors, and may not have the computing facilities to enable him to do this even should he wish to, and therefore an attempt has been made here to present the information in a readily usable form.

Improved formulation of the carry-over factor
As shown in equation 4.35, the carry-over factor is a function only of λl. That equation has therefore been produced in graphical form, Figure 4.23. In order to know the value of the bi-moment carry-over factor for any particular

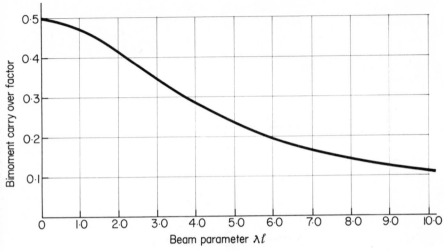

Figure 4.23

problem it is necessary therefore only to calculate the value of λl for the section under investigation. Reference to the graph then enables the carry-over factor to be evaluated.

Improved formulation of the distribution factor
Equations 4.33 and 4.34 show that the moment distribution factors are dependent not only upon the λl values of the members framing into the joint under consideration, but also the end conditions of the members. Since an infinite variety of combinations of λl may be obtained, it is not practicable to provide graphical details of the values of distribution factors to cover all possible combinations of member sizes and lengths. A very limited study is therefore presented, with three standard cases shown in Figure 4.24 being examined. Although the beams shown are two span only, it must be realized that this is not a limitation on the number of spans to which the technique may be applied, but is due to the method of calculating distribution factors which only considers the members framing into the joint under examination. The presence of further spans will only affect the distribution factors in that the end conditions of the members directly under investigation may be affected.

The three cases examined all consider members of equal size spanning over equal distances.

Case 1 Both ends simply supported with respect to warping. For all values of λl the bi-moment distribution factors at the joint will be 0.5.

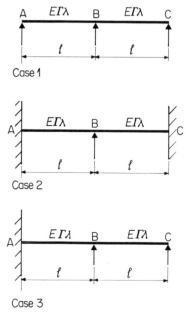

Figure 4.24

Case 2 Both ends fully fixed with respect to warping. For all values of
   λl the bi-moment distribution factors at the joint will be 0.5.

Case 3 One end simply supported with respect to warping, the other end
   fully fixed with respect to warping. The bi-moment distribution
   factors for this case are not constant, but vary with λl. Figure
   4.25 shows the graph of distribution factor against λl for both sides
   of the joint.

Using the information given above, the designer will now be able to evaluate the
bi-moment introduced into a member by the loading and restraint conditions,
and hence calculate the longitudinal stresses that are caused by warping. The
application of this information to the design procedure will now be examined.

   To illustrate the calculations involved in determining these longitudinal
stresses, and the application of the analysis developed in the previous section,
two beam systems will be considered:

(1) A single span beam, simply supported and carrying a uniformly
    distributed load.

(2) A multispan beam carrying a uniformly distributed load.

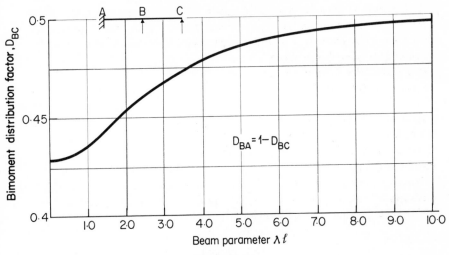

Figure 4.25

Example 4.4
Single span beam. Consider a simply supported beam carrying a load of 1500 N/m, over the full flange width, and spanning 4 m (Figure 4.26). The beam is free to warp but has twisting restraint at the supports. Choose a suitable channel section to carry the load over the required span, taking into account the stresses due to torsion and bending. The yield stress of the material is $\sigma_y = 250$ N/mm^2

$$\text{maximum bending moment} = \frac{wl^2}{8} = 3000 \text{ N m}$$

Taking $\sigma_p = 0.65 \, \sigma_y = 162.5$ N/mm^2, the approximate section modulus required $= 3000 \times 10^3/162.5 = 1.85 \times 10^4$ mm^3. However, this value does not allow for

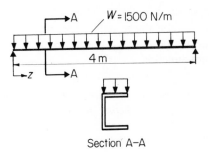

Section A–A

Figure 4.26

any reduction of the section modulus due to the reduced effective width of the compression flange nor for increased stresses due to torsion, hence we start the design proper by choosing a section having a value appreciably greater than the approximate value shown above.

Try a 120 × 50 × 3 channel section ($Z \cong 2.35 \times 10^4$ mm^3). From Figure 2.13 for $b/t = 17$, $C_L = 0.818$, therefore effective width of the compression flange = 40.9 mm, and calculation of the reduced section modulus in compression gives a value of 2.08×10^4 mm^3. The shear centre distance from the web = 17.9 mm (from Table 4.1) and the torque about the shear centre = m = load × ($e + b_f/2$), therefore $m = 64.4$ N mm/mm. The approximate maximum bi-moment for this loading and support condition is $B_{app} = -ml^2/8$ (Table 4.2).

True bi-moment = $B = B_{app} \times F$, where F is the correction factor for this load and support condition and may be determined from Figure A1.2 for the value of the beam parameter λ applicable to this problem.

$$\lambda = \sqrt{\left(\frac{GJ}{E\Gamma}\right)}$$

where E = modulus of elasticity, taken as 200 kN/mm^2
G = the rigidity modulus, taken as 82.6 kN/mm^2
and J and Γ are found for the channel section from Table 4.1.

Hence $\lambda = 5.58$, and from Figure A1.2, the correction factor $F = 0.225$

$$\therefore \text{actual bi-moment } B = 0.225 \times B_{app} = -0.225 \times \frac{m \times l^2}{8}$$

The longitudinal stress due to this bi-moment is obtained from the equation

$$\sigma = \frac{B\omega}{\Gamma}$$

The values of ω for the following positions for the particular section under investigation are:

(a) at the web/flange junction, $\omega = 1075$ mm^2
(b) at the free edge of the flange $\omega = 1926$ mm^2

Therefore the longitudinal stresses due to torsion are:

(a) at the web/flange junction,

$$\sigma_T = \frac{B_{app} \times F \times \omega}{\Gamma} = 74.5 \text{ N/mm}^2 \text{ (tension at upper junction, compression at lower junction)}$$

(b) at the free edge of the flange,

$$\sigma_T = 133 \text{ N/mm}^2 \text{ (compression at edge of upper flange, tension at edge of lower flange)}$$

The maximum longitudinal stress due to bending

$\sigma_B = \dfrac{M}{Z_{red}}$ (where Z_{red} is the reduced section modulus in compression as previously calculated)

$\therefore \sigma_B = 144 \text{ N/mm}^2$

These stress distributions may be shown in diagrammatic form as in Figure 4.27. The total stress distribution is determined from a summation of the bending stresses and the torsional stresses, and as can be seen from Figure 4.27, gives a maximum stress intensity of $144 + 133 = 277 \text{ N/mm}^2$ at the free edge of the flanges

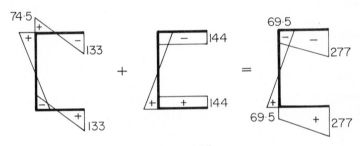

Figure 4.27

The British Standards Specification BS 449 Addendum 1 stipulates that the total stress due to the sum of the effects of bending and torsion must not be greater than the material yield stress. This section is therefore unsuitable.
Try a 100 x 50 x 4 channel section ($Z = 2.23 \times 10^4 \ mm^3$). Using calculations as before, the maximum longitudinal stress due to bending $\sigma_B = 120 \text{ N/mm}^2$, and the maximum longitudinal stress due to torsion $\sigma_T = 64.6 \text{ N/mm}^2$. Therefore the total longitudinal stress $= 120 + 64.6 = 184.6 \text{ N/mm}^2$. This is less than the permitted stress given in BS 449 Addendum 1 and the section is therefore suitable.

Now let us examine the effects on the longitudinal stresses of fixing the ends of the beam so that both warping and twisting are prevented. Considering the bi-moment/bending moment analogy, this is equivalent to providing a built-in end to the beam.

Under this condition of loading, the bending moment will vary between $wl^2/12$ (hogging) at the supports to $wl^2/24$ (sagging) at the centre of the span.

therefore maximum bending moment $= \dfrac{wl^2}{12} = 2000 \text{ N m}$

Taking σ_p as $0.65\sigma_y$, we get $\sigma_p = 162.5$ N/mm^2, and the approximate section modulus required is

$$\frac{2000}{162.5} \times 10^3 = 1.23 \times 10^4 \text{ mm}^3$$

Try a 100 x 50 x 3 channel section ($Z = 1.74 \times 10^4$ mm^3). From Chapter 2, $b/t = 17$, and $C_L = 0.818$, so that the effective width of the compression flange is 40.9 mm and calculation of the reduced section modulus in compression gives a value of 1.65×10^4 mm^3. The shear centre distance from the web = 18.75 mm and therefore the torque about the shear centre = $m = 65.6$ N mm/mm.

From Table 4.2, the approximate bi-moment for this loading and support condition is

(a) at enus, $\dfrac{ml^2}{12}$

(b) at centre, $-\dfrac{ml^2}{24}$

True bi-moment = $B_{app} \times F$, where F is the correction factor which for this type of loading and support may be determined from Figure A1.2.

For this section,

$$\lambda = \sqrt{\left(\frac{GJ}{ET}\right)} = \sqrt{\left(\frac{82.6 \times 1670}{200 \times 219 \times 10^6}\right)} = 0.00177 \text{ mm}^{-1}$$

so that $\lambda l = 7.1$

Therefore from Figure A1.6, the correction factor at the ends is 0.6 and at the mid-span position 0.39, so that the actual bi-moment values are:

(a) at ends,

$$0.6 \times B_{app} = \frac{0.6\, mL^2}{12}$$

(b) at mid-span,

$$0.39 \times B_{app} = -\frac{0.39\, mL^2}{24}$$

Maximum longitudinal torsional stresses will occur either at the free edge of the flange, or at the web/flange junction. The sectorial co-ordinate values (ω) for these two positions are:
(i) at the web/flange junction $\omega = eb_w/2 = 973.5$ mm^2
(ii) at the free edge of the flange $\omega = (b_f - e)\, b_w/2 = 156.5$ mm^2
therefore the longitudinal stresses due to torsion are (refer to Figure 4.28 for signs):

(a) at the supports:

  (i) at the web/flange junction,

$$-\frac{0.6 \times 65.6 \times 16 \times 10^6 \times 973.5}{12 \times 219 \times 10^6} = -233 \text{ N/mm}^2$$

  (ii) at the free edge of the flange,

$$-\frac{0.6 \times 65.6 \times 16 \times 1562.5 \times 10^6}{12 \times 219 \times 10^6} = -374 \text{ N/mm}^2$$

(b) at mid-span:

  (i) at the web/flange junction,

$$\frac{0.39 \times 65.6 \times 16 \times 10^6 \times 973.5}{24 \times 219 \times 10^6} = 75.8 \text{ N/mm}^2$$

  (ii) at the free edge of the flange,

$$\frac{0.39 \times 65.6 \times 16 \times 10^6 \times 1562.5}{24 \times 219 \times 10^6} = 121.7 \text{ N/mm}^2$$

The longitudinal stresses due to bending will be:

(a) at the supports,

$$\frac{2000 \times 10^3}{1.65 \times 10^4} = 121 \text{ N/mm}^2$$

(b) at mid-span,

$$\frac{1000 \times 10^3}{1.65 \times 10^4} = 60.5 \text{ N/mm}^2$$

The stress distribution around the section due to the combination of bending and torsion is as shown in Figure 4.28. At the supports the bending stresses are reversed due to the hogging moment, but it may be seen from Figure 4.28 that the maximum stress is greater than the yield stress at both the web/flange joint, and at the free edge of the flange, and this section is therefore unsuitable.
Try a 120 x 50 x 5 channel section. Using calculations as before, we find that the longitudinal stresses due to torsion are;

(a) at the supports:

  (i) at the web/flange junction, $\sigma_T = 94.7$ N/mm^2
  (ii) at the free edge of the flange, $\sigma_T = 170$ N/mm^2

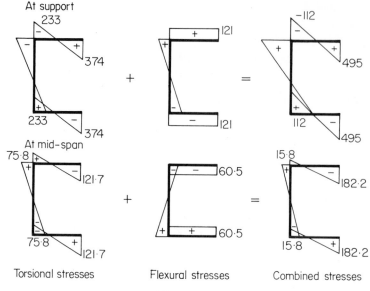

At support

At mid-span

Torsional stresses   Flexural stresses   Combined stresses

Figure 4.28

(b) at mid-span:

    (i) at the web/flange junction, $\sigma_T = 20.3$ N/mm²
    (ii) at the free edge of the flange, $\sigma_T = 36$ N/mm²

The longitudinal stresses due to bending are;
(a) at the supports, 60 N/mm²
(b) at mid-span, 30 N/mm²

    The combination of stresses in this section is shown in Figure 4.29, from which it may be seen that the maximum longitudinal stress occurs at the web/flange joint. It is less than 250 N/mm² (yield stress) and therefore the section size is suitable.

    The above calculations provide extremely good examples of the necessity of including torsional effects in any calculations of longitudinal stresses. If such effects had not been included, in both examples the first section chosen would have been considered adequately strong for the specified problem, but inclusion of the torsional calculation increases the maximum stress in the section to beyond yield value. The examples also illustrate the advantage to be gained by choosing a section size that has a low value of λ (i.e. the ratio of torsion constant to warping constant is increased).

    In calculating the torque about the shear centre, the above solutions assume that the load is uniformly distributed across the full flange width. In many load applications however, any rotation of the beam will permit a redistribution of the load to occur, so that in practice the actual torque may be reduced by virtue of an effective reduction of the torque lever arm due to load redistribution over the flange. Any reduction of torque will, of course, lead to reduced bi-moment, and therefore reduced warping stresses.

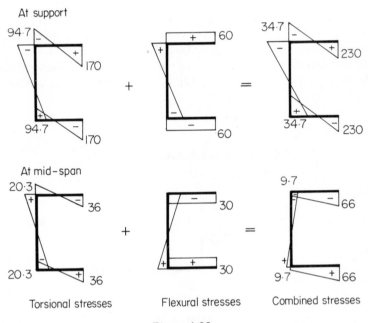

Figure 4.29

Exact assessment of this effect is tedious, and the code of practice makes no allowance for the load redistribution due to twist, but for absolute economy of design it should be considered.

Example 4.5: Multispan beam
It is required to evaluate the safe load that may be carried by the three span beam system shown in Figure 4.30. All spans are equal, the section is constant throughout, and the material and cross-sectional properties required for the calculation are given below.
Material properties

$$E = 200 \times 10^9 \text{ N/m}^2, \quad G = 82.5 \times 10^9 \text{ N/m}^2$$

Cross-sectional properties. The cross-section dimensions are shown in Figure 4.30. From these, and from the information given in Table 4.1, the cross-sectional properties may be calculated to be:

$$Z_x = 20.74 \times 10^3 \text{ mm}^3$$

$$J = 355 \text{ mm}^4$$

$$\Gamma = 74 \times 10^7 \text{ mm}^6$$

The maximum value of the sectorial co-ordinate is at the free edge of the flange, $\omega_{max} = 3360 \text{ mm}^2$. The minimum value of the sectional co-ordinate is at the web/flange junction, $\omega_{min} = 900 \text{ mm}^2$.

The permissible stress in bending is $\sigma_p = 0.65\sigma_y$ where σ_y is the yield stress of the material. If $\sigma_y = 250 \text{ N/mm}^2$, then $\sigma_p = 162.5 \text{ N/mm}^2$. For combined

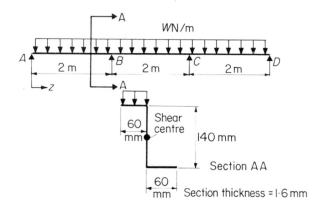

Figure 4.30

bending and torsion the permissible stress is σ_y. The load on the beam is taken as effectively acting over the total flange width. This gives rise to a uniformly distributed bending load of w N/m, and a uniformly distributed torque of m Nm/m, where $m \simeq w \times b_f/2$ (NB both loadings are relative to the shear centre axis).

Maximum bending moment. The maximum bending moment occurs in this case at an internal support and has the value

$$\frac{wl^2}{10} \text{ i.e. } \frac{w \times 4}{10} = 0.4w \text{ N m}$$

Maximum bi-moment. To evaluate the bi-moment at all points of the beam, and hence to obtain the maximum bi-moment in the beam, the procedure described in Section 4.4.2 must be followed.

$$\lambda = \sqrt{\left(\frac{GJ}{ET}\right)} = \sqrt{\left(\frac{82.5 \times 10^9 \times 355}{200 \times 10^9 \times 74 \times 10^7}\right)} = 4.44 \times 10^{-3} \text{ mm}^{-1}$$

$$\therefore \lambda l = 4.44 \times 10^{-3} \times 2 \times 10^3 = 8.88$$

Carry-over factor. From Figure 4.23, the bi-moment carry-over factor for a λl value of 8.88 is 0.128.

Distribution factors. From Figures 4.24 and 4.25, the bi-moment distribution factors at joints B and C for $\lambda l = 8.88$ are:

Joint B, D_{BA} = 0.496
$\qquad\quad D_{BC}$ = 0.504
Joint C, D_{CB} = 0.504
$\qquad\quad D_{CD}$ = 0.496

The initial fixed-end bi-moments (FEBs) are obtained in the same way as for the corresponding case for plane bending (note that in this case the loading applies a negative torque).

$$\text{Fixed-end bi-moment} = B_{BC} = B_{BA} = -\frac{ml^2}{12} = -\frac{m \times 4}{12} = -0.333m \text{ N m}^2$$

but $m \simeq w\dfrac{b}{2} = w \times 30 \text{ N mm/m}$

$\therefore B_{CB} = -0.333 \times w \times 30 \times 10^3 = -10\,000w \text{ N mm}^2$ and $B_{BC} = 10\,000w \text{ N mm}^2$

Bi-moment distribution

	B		C	
d/fs	0.496	0.504	0.504	0.496
FEBs		$-10\,000w$	$10\,000w$	
	4960	5040 $\quad\rightarrow$	645	
		$-688 \quad\leftarrow$	-5370	-5275
	347	347 $\quad\rightarrow$	45	
			-23	-22
	5301w	$-5301w$	5297w	$-5297w$

The bi-moment diagram for the complete structure is shown in Figure 4.31, the value of bi-moment at the centre of the middle span may be calculated from consideration of the conditions pertaining to the corresponding plane bending case, together with the correction factor given in Figure A1.6. From Figure 4.31, it may be seen that the maximum bi-moment occurs at an internal support, and has the value $5300w$ N mm^2

Maximum longitudinal stress. This stress is due to the combined effects of bending and torsional bi-moment and is given by the appropriate algebraic sum of values obtained from the expressions M_x/Z_x and $B\omega/\Gamma$.

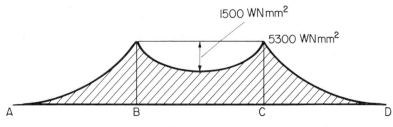

Figure 4.31

In the example under consideration, the maximum stress will occur either at the web/flange junction, or at the free edge of the flange and both positions must be checked:

(1) Top web/flange junction

$$\sigma = -\frac{0.4 \times 10^3 w}{20.74 \times 10^3} + \frac{(900 \times 5300w)}{74 \times 10^7}$$

$$\sigma = -0.0128w \text{ N/mm}^2$$

(2) Top free edge of the flange

$$\sigma = -\frac{0.4 \times 10^3 w}{20.47 \times 10^3} + \frac{(+5300w \times -3360)}{74 \times 10^7}$$

$$\sigma = -0.0433w \text{ N/mm}^2$$

with the negative sign indicating that the stress is compressive. The critical value is (2), i.e. $0.0433w$ N/mm^2, and from $\sigma \not> \sigma_y$, the safe load may be determined.

Thus for the maximum load, $0.0433w = 250$

therefore

$$\underline{w = 5766 \text{ N/m}}$$

If the longitudinal torsional stresses are ignored, the apparent safe load for bending considerations only is found from

$$162.5 = \frac{M_x}{Z_x} = \frac{400w}{20.74 \times 10^3}$$

giving

$$\underline{w = 8450 \text{ N/m}}$$

4.5 REFERENCES

1 Walker, A. C., *Torsion,* Chatto & Windus, 1975.
2 Vlasov, V. Z., 'Thin-walled elastic beams,' Israel Program for Scientific Translations, 1961.
3 Timoshenko, S. P. and Gere, J. M., *Theory of Elastic Stability,* McGraw-Hill, 1961
4 Chilver, A. H., *Proc. ICE,* Vol. 20, p. 233, 1961.
5 Black, M. M. and Semple, H. M., *Int. J. Mech. Sci.,* Vol. 11, p. 791, 1969.
The following texts are recommended for further reading:
Megson, T. H. G., *Analysis of Thin-walled Members,* Intertext, 1975.
Oden, J. T., *Mechanics of Elastic Structures,* McGraw-Hill, 1967.
Zbirohowski-Koscia, K., *Thin-walled Beams, From Theory to Practice,* Crosby Lockwood, 1967.
Terrington, J. S., 'Combined bending and torsion of beams and girders,' BCSA Publication no. 31 (first part).
Bethlehem Steel Corporation, 'Torsion analysis of rolled steel sections.'
Pettersson, O., 'Method of successive approximations for design of continous I beams submitted to torsion', *Publications IABSE,* Vol. 15, 1955.

5

Various Aspects of Detail Design

5.1 LOCAL BUCKLING OF WEBS

If loads are applied locally to a beam, for example if the length of the supports is approximately the same as the depth of the beam, or the load is applied by a cross-member(see Figure 5.1), then high compressive stresses will be induced locally in the web of the beam. These stresses may be of such a magnitude that the web will buckle and failure occur at the position of the reaction or load application. This type of effect, which is called *web local buckling,* or sometimes *web crushing,* is discussed in this section and we begin by considering some analytical and experimental studies into the more general behaviour of plates subject to local edge compressive loading. Later, the design formulations used in the British design specification are presented and finally we discuss some measures that may be taken to counteract the onset of web local buckling.

5.1.1 LOCAL BUCKLING OF A LONG RECTANGULAR PLATE LOADED ALONG PART OF ITS LENGTH

Before discussing the crushing of the web of a section it is instructive to consider the behaviour of a representative plate that is subjected to compressive loading along only part of its edge. Taking first the simplest situation, shown in Figure 5.2, in which we have a rectangular plate with loading along two opposite edges. When the plate is initially perfectly flat the stress distribution is quite complex, with the stresses diminishing in intensity away from the zone of loading. It is not possible to obtain an exact solution for the stresses σ_X, σ_Y, etc, that can be used in the stability equation 2.7 to determine the critical value of the load.

However it is observed during tests that, if all the edges of the plate are simply supported, the plate will buckle into a shape similar to that shown in Figure 5.3. Using a simple expression to describe this buckled shape, together

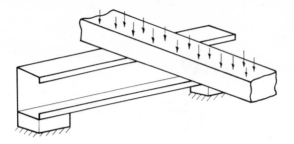

Figure 5.1

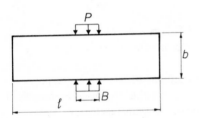

Figure 5.2

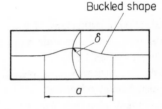

Figure 5.3

with an approximate solution of the stress distribution, the critical load P_{cr} can be calculated from[1]

$$P_{cr} = \frac{k \, \pi^2 \, Et^3}{12 \, (1 - \nu^2) \, b} \tag{5.1}$$

where

$$k = \left(\alpha + \frac{1}{\alpha}\right)^2 \beta\alpha^2 \left[\alpha^4 \left(\frac{\beta}{\alpha} + \frac{1}{\pi} \sin \frac{\pi\beta}{\alpha}\right) - \frac{4}{\pi} \frac{\sin \dfrac{\pi\beta}{\alpha}}{\left[3\left(\dfrac{1}{\alpha}\right)^2 + 2 + 3\,\alpha^2\right]}\right]^{-1} \tag{5.2}$$

and $\alpha \equiv a/b$, $\beta \equiv B/b$. The value of the wavelength a corresponding to a given geometry b, B and l is obtained from the solution of

$$\frac{4\alpha}{1 + \alpha^2} - \frac{3\alpha^2\beta + \dfrac{1}{\pi}\left(4\alpha^3 \sin \dfrac{\pi\beta}{\alpha} - \pi\alpha^2\beta \cos \dfrac{\pi\beta}{\alpha}\right)}{\alpha^3\left(\beta + \dfrac{\alpha}{\pi} \sin \dfrac{\pi\beta}{\alpha}\right)} = 0 \tag{5.3}$$

Typical values for the critical load coefficient k are shown in Figure 5.4. The interesting feature here is that, provided the plate length is more than twice the

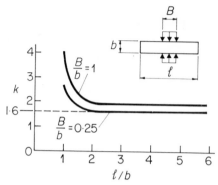

Figure 5.4

length of the loading zone ($l/b > 2$), the coefficient k in effect ceases to be a function of l and for a wide range of loading geometries k can be taken conservatively as 1.6.

Like the uniformly compressed plates considered in Chapter 2, the plates supported on all edges and subject to local edge loading do not collapse at their critical loads. However, the analysis of the post-buckling behaviour is very intricate and to date no theoretical expression corresponding to equation 2.21 has been derived. Tests[1] have been carried out with a variety of loading geometries and Figure 5.5 (a) shows some example results. The relationship

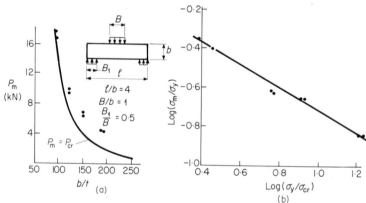

Figure 5.5

between the parameters (P_m/P_y) and (P_{cr}/P_y), which were developed in Chapter 2, is shown in Figure 5.5 (b), from which we can derive the expression

$$\frac{P_m}{P_y} = 0.51 \left(\frac{P_{cr}}{P_y}\right)^{0.47} \tag{5.4}$$

in particular for $l/b = 5.75$, $B/b = 1$. This has a similar form to the expression obtained by Chilver (Reference 22 of Chapter 2) for the buckling of uniformly compressed sections. Equation 5.4 of course includes the effects of initial imperfections in the plate.

Another loading geometry that occurs in practice is that shown in Figure 5.6 (a), that is where a local compressive load is reacted by shear forces at the

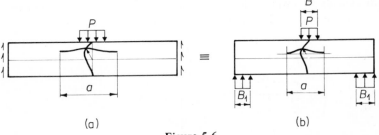

(a)   (b)

Figure 5.6

ends of the plate. This type of loading may be shown[2] to be equivalent, from a buckling point of view for a long plate $l/b \geqslant 4$, to the loading geometry shown in Figure 5.6 (b). The values of the critical load coefficient k are shown in Figure 5.7 (the analytical expressions for k are given in Appendix 3). For this loading case k varies with the length of the plate since buckling is induced by the bending compressive stresses as well as the stresses due directly to the applied loading. The buckled deflection shape is shown in Figure 5.6. Results of some tests are shown in Figure 5.8 (a) and we see once more that the plate can support

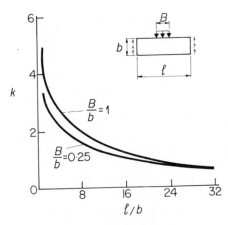

Figure 5.7

loads greater than the critical load without collapsing. Figure 5.8 (b) shows that
the collapse load and the critical load are related by

$$\frac{P_m}{P_y} = 0.72 \left(\frac{P_{cr}}{P_y}\right)^{0.58}$$

(5.5)

in particular for $l/b = 4$, $B/b = 1$. A number of other loading and boundary
conditions have also been studied and are reported in References 1 and 2. Also
the critical loads for a wide variety of plates have been derived using numerical
methods, and results are given in References 3–5.

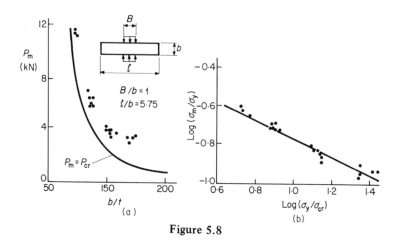

Figure 5.8

5.1.2 DESIGN FORMULATION

Although the preceding analysis indicates the type of behaviour to be expected
in a locally loaded plate, the results cannot be applied immediately to the design
of beam webs. There are several reasons for this, one being that the edges of a
web are not simply supported but, as we saw in Chapter 3, are in fact subject to
rotational restraint from the flanges. These flanges also carry a considerable
portion of the bending stresses so that the critical load of the web will not be
influenced by the length of the beam as much as is indicated in Figure 5.7.
Another reason is that the load is not really applied in the plane of the web;
instead, as shown in Figure 5.9 (a), the bend radii cause the load to be effectively
offset from the web. Further, in the analysis, we have considered that all edges
of the plate were supported, but of course it is not unusual for the web to be
unsupported at the end of a beam (see Figure 5.1). Research is continuing in an
effort to incorporate these practical aspects of cold-rolled sections into the
general plate analysis outlined in Section 5.1.1, but meanwhile for design

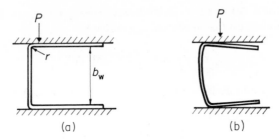

Figure 5.9

purposes we must resort to empirical formulations derived by Winter[6] on the basis of a large number of test results.

These expressions are presented in the American Specification[7] and incorporated in BS 449 Addendum 1, as follows: For beams having a single unreinforced web, with inside corner radius r equal to the sheet thickness, and: (1) For a concentrated load or reaction on the ends of the beam or the outer ends of cantilevers, the permissible maximum stress σ_m (N/mm^2) is

$$\sigma_m = \frac{P_m}{Bt} = \left[31.8 + 742 \left(\frac{t}{b_w}\right)\left(\frac{b_w}{B}\right) - 0.17 \left(\frac{b_w}{t}\right) - 0.08 \left(\frac{b_w}{B}\right)\right]$$

$$\times \left[1.33 - 0.33 \frac{\sigma_y}{250}\right]\frac{\sigma_y}{250} \quad (5.6)$$

For a corner radius exceeding t the value obtained for σ_m from equation 5.6 should be multiplied by

$$1.15 - 0.15 \, r/t$$

(2) For reactions at intermediate supports and for concentrated loads within a span, where the inside radius does not exceed the thickness of the sheet,

$$\sigma_m = \left[17.4 + 2311 \left(\frac{t}{b_w}\right)\left(\frac{b_w}{B}\right) - 0.068 \left(\frac{b_w}{t}\right) - 3.79 \left(\frac{b_w}{B}\right)\right]$$

$$\times \left[1.22 - 0.22 \frac{\sigma_y}{250}\right]\frac{\sigma_y}{250} \quad (5.7)$$

For a corner radius exceeding t the value of σ_m obtained from equation 5.7 should be multiplied by

$$1.06 - 0.6 \, r/t$$

For I beams made from two channel sections connected back to back, or any other similar section providing a high degree of restraint against rotation of the web (see Figure 5.10) we have also from Reference 7, for end reactions or for concentrated loads on the outer ends of cantilevers,

$$\sigma_m = \sigma_y \left[33.6 \left(\frac{t}{B} \right) + 4.23 \sqrt{\left(\frac{t}{B} \right)} \right] \qquad (5.8a)$$

and for reactions of interior supports or for concentrated loads located anywhere on the span

$$\sigma_m = \sigma_y \left[50.5 \left(\frac{t}{B} \right) + 10.95 \sqrt{\left(\frac{t}{B} \right)} \right] \qquad (5.8b)$$

Due to the empirical nature of the above expressions there are a number of constraints on their applicability:

(1) If $r > 4t$, equations 5.6–5.8 do not apply and resort should be made to testing;
(2) If $b_w/t > 150$, $B > b_w$ the expressions do not apply;
(3) $\sigma_m \leqslant 0.50\ \sigma_y$;
(4) For loads located close to ends of beams, equations 5.7 and 5.8b apply, provided that for cantilevers the distance from the free end to the nearest end of bearing, and for a load close to an end support the clear distance from edge of end bearing to nearest edge of load, is larger than $1.5\ b_w$. Otherwise equations 5.6 and 5.8a apply.

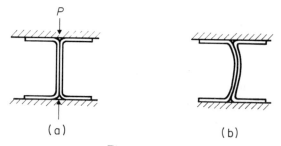

(a)   (b)

Figure 5.10

5.1.3 PREVENTION OF LOCAL BUCKLING

The simplest and most effective way of preventing web local buckling is to provide some form of stiffener at the zone of loading. For example, Figure 5.11 shows a simple stiffener made from a cold-formed angle which is bolted to the web of a channel. If the load is applied through outstands welded, or bolted, to the web (Figure 5.12), then not only is local crushing prevented but also torsion stresses can be largely reduced because the load is applied near the shear centre.

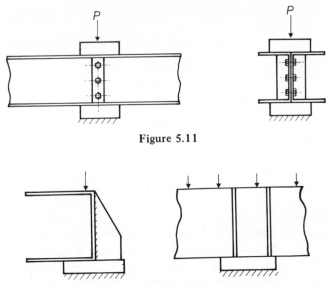

Figure 5.11

Figure 5.12

5.2 BOLTED CONNECTIONS

Because of the intrinsic thinness of cold-formed sections, connections can be a problem and are sometimes critical in design. In the fabrication of cold-formed steel structures, component sections are often bolted together. The connections are required to transmit various types of loading and must be designed to carry these loads without failing.

The loads on connections can generally be divided into two main types:

(1) Loads which tend to shear the bolts, and
(2) Loads which apply tension to the bolts.

The second type of load is generally of less importance than the first, and relatively little is known about its effects. In BS 449 Addendum 1, only shear loads are considered, and appropriate design requirements are specified for connections in shear.

The various types of failure mechanism which can arise in bolted connections in shear can best be explained by considering a simple, one bolt, single lap joint connection, as shown in Figure 5.13. From tests performed in America,[8-10] it has been established that there are four basic failure modes for such a connection.

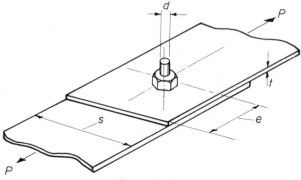

Figure 5.13

These are as follows:
  (1) Longitudinal shearing of the sheet by the bolt as shown in Figure 5.14. This failure type occurs when the edge distance *e* is small.
  (2) Bearing failure, with material shearing and piling up in front of the bolt as shown in Figure 5.15. This is probably the equivalent of (1) for connections with larger edge distance.
  (3) Transverse tension tearing of the sheet as shown in Figure 5.16. This type of failure occurs when the sheet width is small.
  (4) Shearing of bolts.

Failure of a connection in shear can occur in any of these modes, or in a combination of two or more modes.

Examination of these failure mechanisms using purely mathematical means is well nigh impossible in the present state of engineering knowledge, since

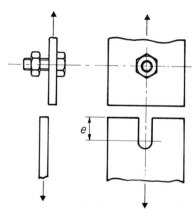

Figure 5.14

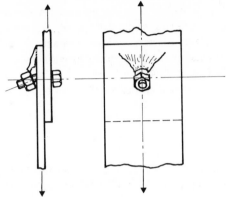

Figure 5.15

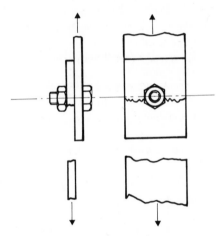

Figure 5.16

failure is affected by a wide variety of additional effects, including bolt clearance, bolt tension, surface finish, material plastic behaviour, crack behaviour, etc. In view of this, empirical rules based on test results, are once more employed to aid the designer.

Winter obtained empirical formulae from his test results governing the first three types of failure. These formulae, which reasonably approximate the failure loads are as follows:

Failure type (1) $P_s = 0.9et\sigma_t$ 　　　　　　　　　　　　　　(5.9)

Failure type (2) $P_b = 3dt\sigma_t$ 　　　　　　　　　　　　　　(5.10)

Failure type (3) $P_t = \left(0.1 + 3\dfrac{d}{s}\right) A_{\text{NET}}\sigma_t$ 　　　　　　　(5.11) where P_s, P_b and P_t are failure loads due to shear, bearing and tension failure respectively, σ_t is the ultimate tensile strength of the material and A_{NET} is the net area of strip measured across a section containing the bolt, i.e. $A_{NET} = (s - d)t$. These formulae, suitably factored, form the basis of both the AISI design specifications and those of BS 449 Addendum 1. In the latter case the design limitations are as follows: The edge distance e shall not be less than $1\frac{1}{2}d$ nor less than

$$\frac{P}{0.4\sigma_t\, t}$$

where P is the applied load. This ensures that the edge distance is more than twice that which would cause shear failure according to equation 5.9 and, therefore, eliminates the possibility of shear failure. Note that in BS 449 Addendum 1 the specification of edge distance is generalized to account for any number of bolts in the line of loading. This is done by stipulating that the edge distance is the distance between hole centres in the line of load, as well as having the definition specified in this section.

The bearing stress on the area $d \times t$ shall not exceed $1.4\sigma_t$,

i.e. $$P_b < 1.4dt\sigma_t$$

This allows a safety factor somewhat greater than 2 on bearing failure according to equation 5.10.

The tensile stress on the net area of section shall not exceed $0.4\sigma_y$ and shall satisfy

$$P_t < \left(0.1 + 3\frac{d}{s}\right) A_{NET}\, 0.4\sigma_t$$

This gives a safety factor of 2.5 on tension failure according to equation 5.11.

With regard to the fourth type of failure, this may be avoided simply by ensuring that the shear stress on the bolt does not exceed the maker's specification.

Now, for any bolted joint on which the loading is known, the satisfaction of the requirements specified above ensures safe design.

At the time of writing, a large scale programme of research is being undertaken into the behaviour of connections in cold-formed sections. Tests are being carried out at various centres in Europe into aspects of bolted, screwed and glued connections, and it is expected that the findings will enable more comprehensive specifications to be introduced for joint design.

The British contribution to this European research programme is sponsored by the CRSA and is concentrated on further research into bolted connections at the University of Strathclyde, Glasgow. From tests already performed at that institution, the accuracy of the design criteria of Winter has been verified for bolted plates, and research is now concentrated on connections between sections rather than sheets. Both shear loadings and tension loadings are under examination.

5.3 DECKING SYSTEMS

A widely used type of cold-formed steel construction is the troughed roof deck as shown in Figure 5.17. Rigorous analyses of such a deck are extremely complex and unsuitable for design purposes. These analyses in general use either folded plate theory or orthotropic plate theory, see for example Reference 11.

From the results of these analyses it has been found that, for design purposes, decks may be considered simply as a series of linked beams, each trough acting

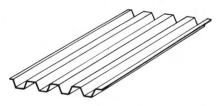

Figure 5.17

as a single beam. Computation of deck strength and stiffness can, therefore, be made using methods similar to those previously discussed for laterally stable beams, and the results obtained will accurately describe the deck behaviour.

At the time of writing, a British Standard Specification[12] dealing with the design of decking systems is in preparation. This uses a modified form of the AISI method for the evaluation of beam strength and deflections, and calculations performed on the basis of Reference 12 give predictions slightly different from those obtained on the basis of BS 449 Addendum 1. In general the results from BS 449 Addendum 1 are more conservative than those of Reference 12, and so design to BS 449 limits a given decking system more severely than a design to Reference 12. A comparison of results obtained from both methods will be obtained later using a simple illustrative example.

5.3.1 LOADINGS
The loadings imposed on roof decks are caused mainly by self-weight of deck with weatherproofing and insulation, by the elements; snow, wind and rain and by loads incurred due to traffic on the roof, either during construction or for

maintenance purposes. Suitable values of these loads for design purposes are tabulated in References 13 and 14. In general, loads due to self-weight, wind, snow, etc. are treated as uniformly distributed loads covering the whole deck, and pose no great problems. Loads incurred during construction or maintenance, however, are generally concentrated; for example the load imposed by a man standing on the roof. This type of loading causes problems in design, since if a load is concentrated at a point on a single trough then it is a severe assumption to imagine that that trough alone must support the load. In reality adjacent troughs deform and help support the applied load, but to estimate the degree of support thus obtained is not an easy matter, requiring once more the sophisticated mathematical methods previously mentioned.

In the proposed new British Standard[12] which is based largely on an existing design method,[15] the problems of concentrated loads are overcome by replacing the specified concentrated loads by empirically derived line loads across the deck, as shown in Figure 5.18, which produce approximately the same magnitude of maximum bending moment. By using this device the analysis can once more proceed on the basis of considering the deck as a series of beams.

The basic loads, other than wind loads, to be considered in design are tabulated in Reference 13 and are as follows.

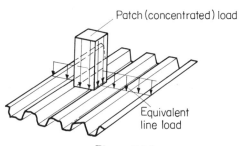

Figure 5.18

Flat roofs

On flats roofs and sloping roofs up to and including 10°, where access (in addition to that necessary for cleaning and repair) is provided to the roof, allowance shall be made for an imposed load, including snow, of 1.5 kN/m², measured on plan, or a load of 1.8 kN concentrated on a square with a 300 mm side.

On flat roofs and sloping roofs up to and including 10°, where no access is provided to the roof (other than for cleaning and repair) allowance shall be made for an imposed load, including snow, of 0.75 kN/m² measured on plan or a load of 0.9 kN concentrated on a square with 300 mm side.

Sloping roofs
On roofs with a slope greater than 10°, and with no access other than for cleaning and repair, the following imposed loads, including snow, shall be allowed for:

(1) for a roof slope of 30° or less: 0.75 kN/m² measured on plan or a vertical load of 0.9 kN concentrated on a square with a 300 mm side.
(2) for a roof slope of 75° or more: no allowance necessary.

For roof slopes between 30° and 75°, the imposed load to be allowed for may be obtained by linear interpolation between 0.75 kN/m² for a 30° roof slope and nil for a 75° roof slope. In addition, loads incidental to construction, which are not covered by Reference 13, should be taken into account.

In Reference 12 the concentrated loads of 0.9 kN and 1.8 kN are replaced by line loads of 1.5 and 3 kN/metre width of deck, as shown in Figure 5.18. Loads incidental to construction are covered by a line load of 2 kN/metre width for design purposes.

Note that since the construction loading mentioned is less than the line load corresponding to the concentrated load case for roofs with access, then this is automatically satisfied if the concentrated load requirement is satisfied. However, for roofs with no access, the construction loading is greater than the concentrated load case, and must be investigated.

5.3.2 EVALUATION OF DECK STRENGTH
The basic method of approach is to consider a single trough, evaluate the second moment of area and section moduli, and from these values determine the load carrying capacity and resistance to deflection of the deck. Since the most widely used type of decking is made up from a number of troughs of the form shown in Figure 5.19 it is advantageous to obtain general expressions for the properties of this type of trough.

Consider, then, the trough shown in Figure 5.19. Using the methods shown in Chapter 3 we tabulate the properties as follows, neglecting the effects of radii and using mid-line dimensions.

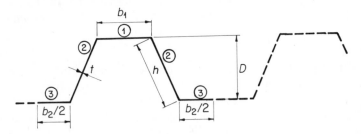

Figure 5.19

Element	b	y	by	$b_y{}^2$	I_{cg}/t	I_{oo}/t
①	b_1	0	0	0	0	0
$2 \times$②	$2h$	$D/2$	hD	$hD^2/2$	$2 \times \dfrac{hD^2}{12}$	$\dfrac{2}{3}hD^2$
$2 \times$③	b_2	D	$b_2 D$	$b_2 D^2$	0	$b_2 D^2$

$$\Sigma b = S = b_1 + b_2 + 2h \qquad \Sigma by = D(b_2 + h) \qquad \Sigma I_{oo}/t = D^2\left(b_2 + \frac{2}{3}h\right)$$

$$\bar{y} = \frac{\Sigma by}{\Sigma b} = \frac{D(b_2 + h)}{S}$$

$$I_{NA} = t\left(\frac{I_{oo}}{t} - \Sigma b\,\bar{y}^2\right) = t\left\{D^2\left(b_2 + \frac{2}{3}h\right) - S\frac{D^2(b_2+h)^2}{S^2}\right\}$$

Expanding S, and simplifying, gives

$$I_{NA} = \frac{tD^2}{S}\left(b_1 b_2 + \frac{2}{3}hS - h^2\right) \tag{5.12}$$

Also, assuming that element ① is in compression

$$Z_c = \frac{I_{NA}}{\bar{y}} = \frac{tD\left(b_1 b_2 + \frac{2}{3}hS - h^2\right)}{b_2 + h} \tag{5.13}$$

Similarly

$$Z_T = \frac{tD\left(b_1 b_2 + \frac{2}{3}hs - h^2\right)}{b_1 + h} \tag{5.14}$$

The section properties can, therefore, be obtained from the above equations. If the loads are such that the compression flange is partly ineffective then b and S are replaced by the effective width and effective developed width, b_e and S_e in the expressions. The use of these expressions is now shown by an example.

Example 5.1
Consider a deck consisting of troughs as shown in Figure 5.20. It is required to obtain the maximum permissible span to carry a load of 1.5 kN/m² measured on plan. Assuming the deck to be continuous over two spans the analysis proceeds as follows. The distributed load for a single trough is obtained by evaluating the trough plan area per metre run, which is 0.125 x 1 = 0.125 m². The load per metre run of trough is, therefore, 0.125 x 1.5 = 0.1875 kN/m. The maximum

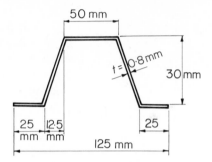

Figure 5.20

moment due to this uniformly distributed loading is, for either a single span or double span beam,

$$M = \frac{wl^2}{8} = \frac{0.1875}{8} \, l^2 = 0.0234 \, l^2 \text{ kilonewton-metres}$$

The trough must therefore be able to withstand safely a bending moment of $23.4 \times 10^3 l^2$ newton-millimetres, where l is the span in metres.

Now $M = \sigma Z$ so that the span should be such that

$$Z_{\min} = \frac{23.4 \times 10^3}{\sigma_A} \, l^2 \text{ mm}^3.$$

If σ_A is taken as 150 N/mm^2 this becomes $Z_{\min} = 156 \times l^2$.

We can now evaluate $Z_{\min}$ and from this obtain l. The value we obtain for $Z_{\min}$ is dependent upon whether we use the method of BS 449 Addendum 1 or the method of Reference 12. We shall firstly use the former method.

From Chapter 2

for
$$\frac{b}{t} = \frac{50}{0.8} = 62.5, \ C_L = 0.642$$

Therefore $b_e = 32.1$ mm

Now $h = \sqrt{(12.5^2 + 30^2)} = 32.5$ mm

and $S_e = 2 \times 25 + 2 \times 32.5 + 32.1 = 147.1$ mm

Using these values in equation 5.13 gives Z_c, which is the lower section modulus in this case (since b_2 is greater than b_e).

This is

$$Z_c = \frac{t D\left(b_1 b_2 + \frac{2}{3} hS - h^2\right)}{b_2 + h}$$

$$= \frac{0.8 \times 30\left(32.1 \times 50 + \frac{2}{3} \times 32.5 \times 147.1 - 32.5^2\right)}{50 + 32.5}$$

i.e. $Z_c = 1088$ mm^3

Now since

$$Z_{min} = 156\, l^2 \text{ we have}$$

$$156\, l^2 = 1088$$

giving $l = \sqrt{\dfrac{1088}{156}} = 2.64$ m

Using the method of Reference 12 we have the following expression for b_e:

$$\frac{b_e}{t} = \frac{700}{\sqrt{\sigma_A}}\left(1 - \frac{700}{4\left(\frac{b}{t}\right)\sqrt{\sigma_A}}\right) \tag{5.15}$$

For $(b/t) = 62.5$ and $\sigma_A = 150$ this becomes

$$\frac{b_e}{t} = \frac{700}{\sqrt{150}}\left(1 - \frac{700}{4 \times 62.5\sqrt{150}}\right) = 44.1$$

Therefore

$$b_e = 35.3 \text{ mm}$$

Hence

$$S_e = 2 \times 25 + 2 \times 32.5 + 35.3 = 150.3$$

Using these values in equation 5.13 gives

$$Z_c = \frac{0.8 \times 30\left(35.3 \times 50 + \frac{2}{3} \times 32.5 \times 150.3 - 32.5^2\right)}{50 + 32.5} = 1153 \text{ mm}^3$$

Hence we obtain

$$l = \sqrt{\frac{1153}{156}} = 2.72 \text{ m.}$$

Note that the effective width from equation 5.15 was higher than that using the C_L values of BS 449 Addendum 1. This is due to the edge support conditions for the troughed sheeting flange being taken into account in Reference 12 and not in

BS 449 Addendum 1. This latter document is not specifically intended for the design of decking. Nevertheless the differences between the two values obtained for Z_c are quite small — about 6%.

As an indication of the variation in effective width obtained using either method, a graph of effective width versus full width is shown in Figure 5.21. The curve A (Reference 12) is drawn for the design stress. This shows that expression 5.15 gives higher effective widths than are obtained from BS 449 over the high b/t range.

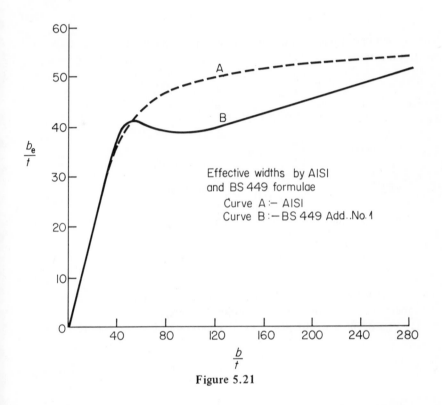

Figure 5.21

It should be noted that the method shown in Reference 12 is not always so easily applied as in this example. If the geometry of the deck is such that Z_T is less than Z_c, then the maximum stress will be tensile. In such a case the effective width of the compression flange cannot be obtained directly since this depends on the stress on the compression flange which in turn depends on the effective width of the compression flange due to its effect on section properties. For this type of problem a trial and error iterative type solution must be employed— involving the need to guess stresses on the compression flange. This will not be discussed here and for further information the reader is referred to Reference 12 (and Reference 16 from Chapter 2).

In general, the deck is required to withstand concentrated loads, as specified by Reference 13. A check is, therefore, necessary to ensure that the deck is sufficiently strong with regard to a concentrated load of 1.8 kN. Using the equivalent line load specified in Reference 12, i.e. 3 kN/metre, the required section modulus can be evaluated.

For one trough the total line load becomes

$$3 \text{ kN/metre} \times 0.125 = 0.375 \text{ kN}$$

The maximum possible moment due to this load applied at the position to give the worst condition is given by Reference 12 as

$$M = \frac{Pl}{4} \times 0.83 = 0.2075 \times 0.375l$$

i.e. $$M = 0.0778l \text{ kN m}$$

From $$\sigma = \frac{M}{Z} \text{ we have}$$

$$Z = \frac{0.0778l \times 10^6}{150} = 518.8l \text{ mm}^3$$

Thus, for the Z value of 1088 mm^3 obtained using the BS 449 method,

$$l = \frac{z}{518.8} = 2.10 \text{ metres}$$

Therefore, the maximum span allowable must be reduced from 2.64 metres to 2.10 metres to satisfy the concentrated load requirement.

The equivalent span obtained using the method of Reference 12 is

$$l = \frac{1153}{518.8} = 2.22 \text{ metres}$$

5.3.3 DEFLECTIONS
Another design requirement is the specification of maximum deflection. The span must not be such that the deflection exceeds stipulated limits. Reference 12 specifies that the maximum deflection based on the uniformly distributed loads should not exceed $l/250$. For a double spanning beam the theoretically correct deflection expression due to a uniformly distributed load is

$$\Delta = \frac{1}{185} \frac{wl^4}{EI}$$

Reference 12 however suggests a more conservative expression to take account of non-uniform loading, etc. This expression is

$$\Delta = \frac{3}{384} \frac{wl^4}{EI} \qquad (5.16)$$

The determination of minimum span from a deflection criterion is rather difficult using the deflection analysis of BS 449 Addendum 1. This is so because M, Δ_{cr} and Δ_a must be obtained before calculation of the actual deflection Δ is possible. Since these quantities are all dependent on the span to different degrees, the resulting equation obtained for l is only suitable for solution by trial and error. These problems are avoided by using the method of Reference 12. This method is an adaptation of the AISI method[7] and assumes that the effective widths of the compression flanges may be obtained from a simple expression which involves only the real width and thickness. This expression is

$$b_e = 100 \, t \left(1 - 25\frac{t}{b}\right) \qquad (5.17)$$

For the flange under consideration this gives

$$b_e = 0.8 \times 100 \left(1 - \frac{25}{62.5}\right) = 48 \text{ mm}$$

It should be pointed out here that in the writer's opinion equation 5.17 is not conservative and overestimates the effective width of the compression flange, for flanges with high b/t ratios, although its effects are counteracted by the various factors of safety in the design approach.

The developed width of the trough is then equal to

$$S_e = 50 + 48 + 2 \times 32.5 = 163 \text{ mm}$$

substituting in equation 5.12 gives

$$I_{NA} = \frac{0.8 \times 30^2}{163} \left(48 \times 50 + \frac{2}{3} \times 32.5 \times 163 - 32.5^2\right) = 21550 \text{ mm}^4$$

Now, if $\Delta = 3/384 \times wl^4/El = l/250$, the minimum span is then

$$l = \sqrt[3]{\left(\frac{1}{250} \times \frac{384 \times 200 \times 10^3 \times 21550}{3 \times .1875}\right)} \times 10^{-3} \text{ m}$$

$$= 2.28 \text{ m}$$

In the example shown the span was taken as the variable to be used for satisfaction of the design criteria. This, of course, is not necessarily the procedure to be adopted in all circumstances. In many cases it may be more useful to change deck shape and dimensions to obtain satisfactory design for a given span.

The equations given in this section are sufficient to permit this to be carried out in a similar fashion as in the illustrative example.

Some general points which may be noted regarding the design of decking are as follows. In most circumstances the most economical deck which can be obtained is that manufactured from the thinnest gauge; the lighter the gauge the greater in theory the strength/weight ratio that can be achieved but there are practical limitations on the useable plate thinness.

The greatest geometrical factor governing the strength and stiffness of a trough is the depth D. Changes in D can alter the second moment of area and section modulus to a great extent with relatively little change in material usage. In comparison, changes in flange widths are relatively insignificant in causing variations of strength and stiffness if the overall trough width is kept constant. Within reason the troughs should be as wide as possible, and to cover a large area economically the webs should be inclined to help widen the trough, although if the web inclination becomes too pronounced (say greater than 45°) other problems arise.

In general, if the ratio of span to depth l/D is large, deflections will govern design; if l/D is small, stresses will govern. Also for small l/D the stresses arising from the concentrated load condition will tend to be most severe. These statements are, of course, only general indications, and each case must be treated individually.

5.4 PURLIN DESIGN

Cold-formed steel purlins behave essentially as do other flexural members, but because of their widespread use and interaction with sheeting a clause in BS 449 Addendum 1 is devoted exclusively to purlin design. The first part of this clause, 122(a), deals with purlins of general shape, and is basically informative. It indicates that, where a purlin is continuous over one or more supports, due allowance may be made for the strength and stiffness of connections. A purlin may also be designed on an experimental basis. Making such allowances can increase the allowable design loads, or reduce the resulting displacements, due to the restraint effect of end connections and the stiffening effect of the roof sheeting.

An example of this is shown in the results given below from a test on a full-scale roof assembly,[16,17] consisting of hot-rolled roof trusses, cold-formed Z-section purlins and corrugated steel sheeting.

For this roof with a fixed slope (21° 48'), the restraining effect of standard end connections (single bolt) (Figure 5.22 (a)) on a single span purlin reduces the theoretical bending moment from $wl/8$ to $wl/9.05$. The normal type of sleeved end connection (Figure 5.22 (b)) reduces it further to $wl/9.87$.

The stiffening effect of roof sheeting, using normal hook bolts (Figure 5.23),

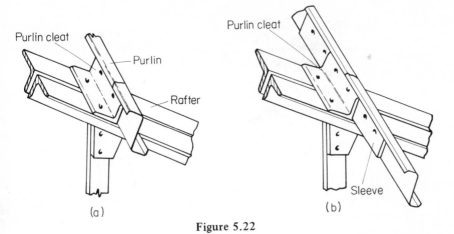

Figure 5.22

on the fixed slope roof further reduces the theoretical bending moment value to $wl/10.6$ for a single span internal purlin fitted with standard end connections and to $wl/11.6$ for sleeved end connections.

The purlin section on the fixed slope roof did not bend about its theoretical neutral axis, but rather about a horizontal axis. Roof slope has little effect on the direction of purlin deflections in that the purlin deflects almost vertically regardless of the roof slope. Consequently it is considered that instead of using the theoretical value of the inclination of the neutral axis, a value of the roof slope should be employed in all computations. Roof sheeting and end connections almost entirely eliminate the twist at normal roof slopes, although there is a small amount of twist at low values of slope.

It is required that connections between purlins and roofing material shall be such as to provide effectively continuous lateral support to the purlins. This rules out the problem of lateral instability, and the stresses prescribed in Table 1 of the specification may, therefore, be ignored. Further general information on the interaction between sheeting and linear members attached by various fastenings is given in Reference 18. Clause 122b of the specification concerns Z purlins, which have been in common use for many years.

For roof slopes not exceeding $30°$, empirical rules are given for the design of Z purlins; the flanges must be square to the webs and have stiffening lips; the lip dimensions must be such that, while adequate support is given to the flange edge, the lips do not themselves, face the possibility of local buckling. Also the width/thickness ratio for the compression flange must not exceed $30\sqrt{(250/\sigma_y)}$. This rigorous limitation of flange and lip dimensions ensures that local buckling is kept to a minimum, and that in most instances the compression

156   COLD-FORMED SECTIONS

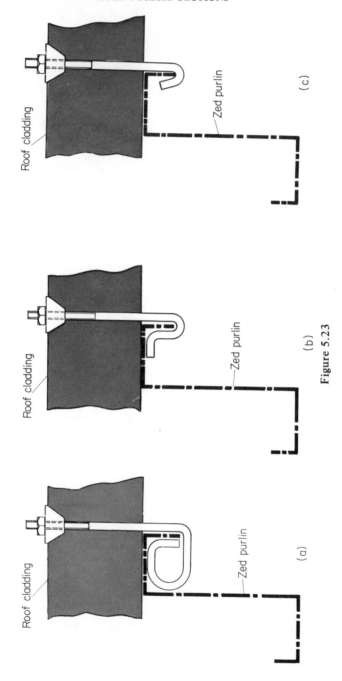

Figure 5.23

flange is fully effective. (In reality, these specifications were designed originally to eliminate local buckling completely, since the specification of lip dimensions ensured that the flange could be treated as a stiffened plate. Using the C_L factors in vogue at the time, the original specification was made up. A stiffened plate with b/t equal to $30\sqrt{(250/\sigma_y)}$ had a C_L factor of 1, and was thus fully effective. Using the present values of C_L, a flange with a b/t ratio of 30 is not quite fully effective, but has a C_L value of 0.99. Thus the orginal aim of these specifications is marginally undermined, although they still perform the function of keeping the flanges very effective.)

Clause 122b also lays down minimum values for section dimensions and properties in terms of the purlin span length l in millimetres. The depth D of the section must not be less than $l/45$, the total width over both flanges must not be less than $l/60$, and the numerical value of the section modulus Z must not be less than $wl/1.8 \times 10^{-3}$ cm^3 where w is the total distributed load on the purlin in kilonewtons.

The limitation on Z imposes an upper bound on the maximum stress. The magnitude of maximum stress obtained may be evaluated using the expression $\sigma = M/Z$. Neglecting any effects of restraint due to connections, M_{max} is equal to $wl/8$. Therefore.

$$\sigma_{max} = \frac{wl \times 10^3}{8(Z \times 10^3)} \text{ N/mm}^2$$

If

$$Z = \frac{wl}{1.8} \times 10^{-3} \text{ cm}^3$$

then

$$\sigma_{max} = \frac{wl}{8} \times \frac{1.8}{wl \times 10^{-3}} = 225 \text{ N/mm}^2$$

Therefore, regardless of the tensile strength of material used, the limitations of Clause 122b rule out any possiblity of designing for stresses greater than 225 N/mm^2.

The limitations on width and depth of the section do not perform such a straightforward function as that performed by the limitation on Z. Indeed, these limitations were originally taken straight from the parent specification, BS 449, where they were applied to angle section purlins. In general, limiting the depth to not less than $l/45$ imposes a certain maximum value of deflection on the purlin when used in conjunction with the Z limitation, and stating a minimum value of $l/60$ for the total flange width imposes a minimum value of l/r_Y for the purlin, thus guarding against lateral instability.

5.5 REFERENCES

1   Khan, M. Z. and Walker, A. C., *The Structural Engineer,* Vol. 50, 1972.
2   Khan, M. Z., Ph.D. Thesis, University of London, 1972.
3   Zetlin, L., *Proc. ASCE, Eng. Mech. Div.,* Vol. 81, p. 795, 1955.
4   White, R. N. and Cottingham, W. S., *Proc. ASCE, Eng. Mech, Div.,* EM 5, p. 88, 1962.
5   Rockey, K. C. and Bagchi, D. K., *Int. J. Mech. Sci.,* Vol. 12, p. 61, 1970.
6   Winter, G. and Pian, R. H. J., Bulletin No. 35/1, Cornell University Engineering Experimental Station, Ithaca, 1956.
7   'Specification for the design of cold-formed steel structural members', AISI, 1968.
8   Winter, G., *Proc. ASCE,* Vol. 82, Paper No. 920, 1956
9   Winter, G., *Publications IABSE,* Vol. 16, 1956.
10   Dahalla, *Proc. ASCE,* Vol. 97. 1971.
11   Kinloch, H. and Harvey, J. M., *Thin Walled Structures,* Chatto & Windus, 1967.
12   Draft British Standard Recommendations for the Design of Light Gauge Metal Roof Decking, July, 1973.
13   British Standard Code of Practice, CP3: Chapter V: Part 1: 1967.
14   British Standard Code of Practice, CP3: Chapter V: Part 2: 1970.
15   Code of Design and Technical Requirements for Light Gauge Metal Roof Decks, The Metal Roof Deck Association, October 1965.
16   Harvey, J. M., *Thin Walled Steel Structures,* Crosby Lockwood, p. 275.
17   Harvey, J. M., *The Structural Engineer,* Vol. 48. No. 9, September 1970, p. 361.
18   Bryan, E. R., *The Stressed Skin Design of Steel Buildings,* Constrado Monograph, Crosby Lockwood Staples, London, 1972.

Appendix 1

Correction Graphs for Warping Torsion Stresses and Deflections

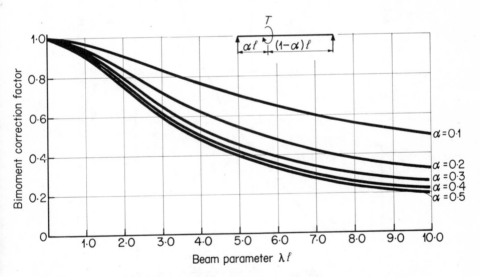

Figure A1.1 Graph of bi-moment correction factor against beam parameter λl; for bi-moment at the load point of the system shown for a simply supported beam.

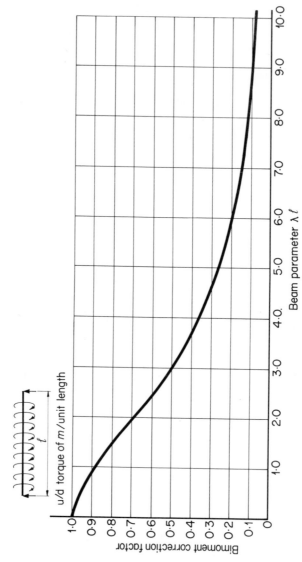

Figure A1.2 Graph of bi-moment correction factor against beam parameter λl; for the mid-point of a simply supported beam under a uniformly applied torque.

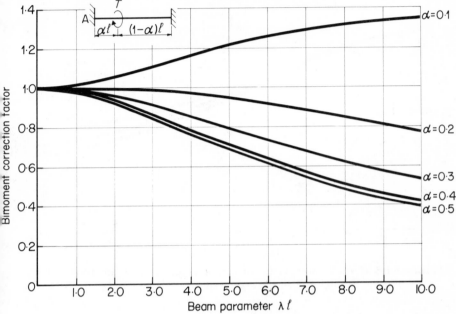

Figure A1.3 Graph of bi-moment correction factor against beam parameter λl; for bi-moment at the load point of the system shown for a built-in beam.

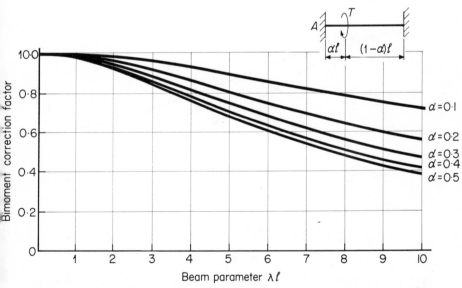

Figure A1.4 Graph of bi-moment correction factor against beam parameter λl; for bi-moment at support A of the system shown for a built-in beam.

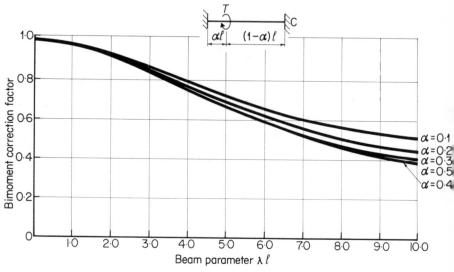

Figure A1.5 Graph of bi-moment correction factor against beam parameter λl; for bi-moment at support C of the system shown for a built-in beam.

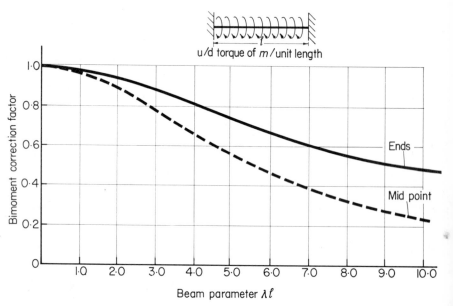

Figure A1.6 Graph of bi-moment correction factor against beam parameter λl; for the ends and the mid-point of a built-in beam under a uniformly applied torque.

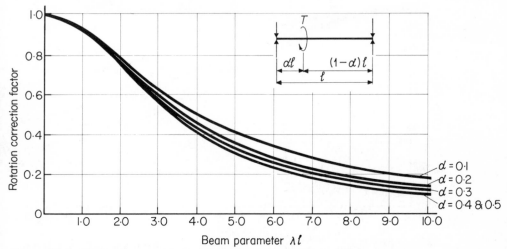

Figure A1.7 Graph of rotation correction factor against beam parameter λl; for rotation at the load point of the system shown for a simply supported beam.

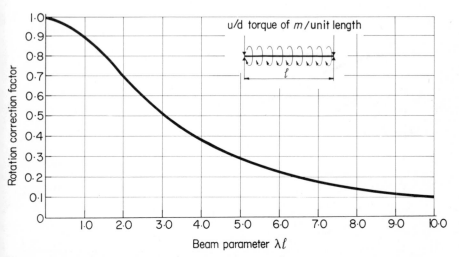

Figure A1.8 Graph of rotation correction factor against beam parameter λl; for rotation of the mid-point of a simply-supported beam under a uniformly applied torque.

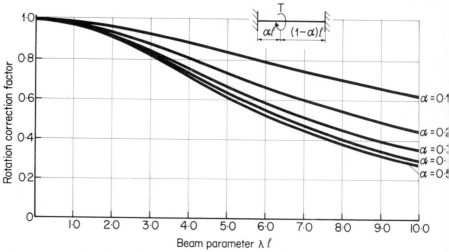

Figure A1.9 Graph of rotation correction factor against beam parameter λl; for rotation at the load point of the system shown for a built-in beam.

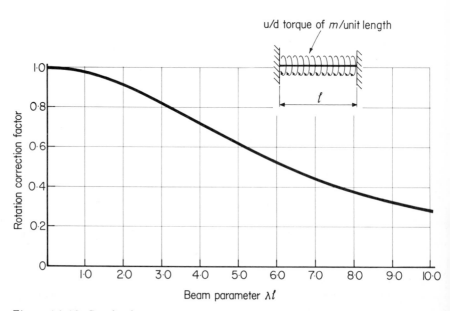

Figure A1.10 Graph of rotation correction factor against beam parameter λl; for rotation of the mid-point of a built-in beam under a uniformly applied torque.

Appendix 2

Formal Solution for Three Span Beam

$$\phi_1 = A \left[\frac{\sinh \lambda_a (z + l)}{\sinh \lambda_a l} - \frac{l + z}{l} \right] \tag{A2.1}$$

$$\phi_2 = \frac{A}{\lambda_b l - \sinh \lambda_b l} \left[\frac{\lambda_a^2}{\lambda_b^2} \sinh \lambda_b (z - l) + \frac{\lambda_a^2}{\lambda_b} l \cosh \lambda_b z + \lambda_a l \coth \lambda_a l \sinh \lambda_b z \right.$$

$$- \sinh \lambda_b z - \frac{\lambda_a^2}{\lambda_b^2} \sinh \lambda_b z - \frac{\lambda_a^2}{\lambda_b^2} (\lambda_b l - \sinh \lambda_b l) + \frac{z}{l} \left(\frac{\lambda_a^2}{\lambda_b^2} l + \sinh \lambda_b l \right)$$

$$\left. - \lambda_a \ l \sinh \lambda_b l \coth \lambda_a l \ - \frac{\lambda_a^2}{\lambda_b} l \cosh \lambda_b l \right) \right]$$

$$+ \frac{T}{E_b \Gamma_b \lambda_b^3} \frac{(1 - \alpha) \lambda_b l - \sinh (1 - \alpha) \lambda_b l}{\lambda_b l - \sinh \lambda_b l} (\sinh \lambda_b z - \lambda_b z) \tag{A2.2}$$

$$\phi_3 = \phi_2 + \frac{T}{E_b \Gamma_b \lambda_b^3} [\lambda_b (z - \alpha l) + \sinh \lambda_b (z - \alpha)] \tag{A2.3}$$

$$\phi_4 = c \left[\frac{\sinh \lambda_a (2l - z)}{\sinh 2\lambda_a l} + \frac{z - 2l}{2l \cosh \lambda_a l} \right] \tag{A2.4}$$

where

$$c = \frac{2A \cosh \lambda_a l}{\mu (\lambda_b l - \sinh \lambda_b l)} \left[\mu \lambda_b l \cosh \lambda_b l + \left(\frac{\lambda_b^2}{\lambda_a^2} l \coth \lambda_a l - \frac{\lambda_b^2}{\lambda_a^2} - \mu \right) \sinh \lambda_b l \right]$$

$$+ \frac{2TL \cosh \lambda_a l (1 - \alpha) \sinh \lambda_b L - \sinh (1 - \alpha) \lambda_b l}{E_b \Gamma_b \mu \lambda_a^2} \frac{}{\lambda_b L - \sinh \lambda_b l}, \tag{A2.5}$$

165

and

$$\frac{A}{\mu l \sinh \lambda_a l} \left[\frac{2\mu l}{\lambda_b} (\lambda_b^2 + \mu \lambda_a^2) \cosh \lambda_b l \sinh \lambda_a l + \frac{2l}{\lambda_a} (\lambda_b^2 + \mu \lambda_a^2) \cosh \lambda_a l \sinh \lambda_b l \right.$$

$$- \left(\frac{\lambda_b^2}{\lambda_a^2} + 2\mu + \mu^2 \lambda_a^2 l^2 \right) \sinh \lambda_b l \sinh \lambda_a l + 2\mu \lambda_b \lambda_a l^2 \cosh \lambda_b l \cosh \lambda_a l$$

$$\left. - \frac{2\mu^2 \lambda_a^2}{\lambda_b} \sinh \lambda_a l - \lambda_b^2 l^2 \cosh \lambda_a l \frac{\sinh \lambda_b l}{\sinh \lambda_a l} \right] = \frac{T}{E_b \Gamma_b \lambda_b^2} [\sinh \alpha \lambda_b l$$

$$+ (1 - \alpha) \lambda_b l \cosh \lambda_b l + \sinh (1 - \alpha) \lambda_b l - \sinh \lambda_b l + \alpha \lambda_b l$$

$$- \lambda_b l \cosh (1 - \alpha) \lambda_b l] \quad - \frac{T}{E_b \Gamma_b \mu \lambda_a^2} (\sinh \lambda_a l - \lambda_a l \cosh \lambda_a l)$$

$$\times [(1 - \alpha) \sinh \lambda_b l - \sinh (1 - \alpha) \lambda_b l] \tag{A2.6}$$

In equations A2.5 and A2.6 $\mu \doteq E_a \Gamma_a / E_b \Gamma_b$. For the case of a uniform-section continuous beam, the elastic constants and section properties will not differ from portion to portion of the beam, thus

$$\lambda_a = \lambda_b; \Gamma_a = \Gamma_b \text{ and } E_a = E_b \tag{A2.7}$$

The equations for this form of the beam can be found by using the relationships A2.6. The solutions are given by equations A2.1–A2.4 where

$$c = \frac{4A \cosh \lambda l}{\lambda l - \sinh \lambda l} (\lambda l \cosh \lambda l - \sinh \lambda l)$$

$$+ \frac{2TL \cosh \lambda l}{E \Gamma \lambda^2} \left[\frac{(1 - \alpha) \sinh \lambda l - \sinh (1 - \alpha) \lambda l}{\lambda l - \sinh \lambda l} \right] \tag{A2.8}$$

$$A \left\{ 8 \cosh \lambda l - 2 - \frac{2\lambda l \cosh 2\lambda l}{\sinh \lambda l} - 3 \frac{\sinh \lambda l}{\lambda l} - \frac{\lambda l}{\sinh \lambda l} \right\}$$

$$= \frac{T}{E \Gamma \lambda^3} \left[\sinh \alpha \lambda l - \frac{\lambda l \sinh (2 - \alpha) \lambda l}{\sinh \lambda l} + 2 \sinh (1 - \alpha) \lambda l \right] \tag{A2.9}$$

Appendix 3

Expression for Critical Load Coefficient

$$k = \cfrac{\beta\left(\alpha^4 + \frac{5\alpha^2}{4} + \frac{17}{32}\right)}{\alpha + \left(a_1\dfrac{\beta}{\alpha} + a_2\dfrac{m\beta}{\alpha^2} - a_3\dfrac{\beta^3}{\alpha^3} + \dfrac{a_4}{\pi}\sin\dfrac{\pi\beta}{\alpha}\right) - \dfrac{9}{8\pi}\cfrac{\sin\dfrac{\pi\beta}{\alpha}}{\left(\dfrac{3}{\alpha^4} + \dfrac{2}{\alpha^2} + 3\right)}}$$

where

$$a_1 = \frac{364}{27\pi^4} - \frac{2}{3\pi^2} + \frac{5}{16}$$

$$a_2 = \frac{4}{3\pi^2} \;;\; a_3 = \frac{2}{9\pi^2}$$

$$a_4 = \frac{364}{27\pi^4} + \frac{5}{16}$$

$$\cfrac{4\alpha^3 + \dfrac{5\alpha}{2}}{\alpha^4 + \dfrac{5\alpha^2}{4} + \dfrac{17}{32}}$$

$$-\cfrac{3a_1\,\alpha^2\beta + 2a_2\,m\,\alpha\,\beta - a_3\,\beta^3 + \dfrac{a_4}{\pi}\left(-\pi\,\alpha^2\beta\cos\dfrac{\pi\beta}{\alpha} + 4\,\alpha^3\sin\dfrac{\pi\beta}{\alpha}\right)}{\alpha^4\left(a_1\dfrac{\beta}{\alpha} + a_2\dfrac{m\beta}{\alpha^2} - a_3\dfrac{\beta^3}{\alpha^3} + \dfrac{a_4}{\pi}\sin\dfrac{\pi\beta}{\alpha}\right)}$$

and $\alpha \equiv \dfrac{a}{b}$; $\beta = \dfrac{B}{b}$; $m \equiv \dfrac{l}{b}$

Subject Index

Author Index

173